Advanced Communication and Control Methods for Future Smartgrids

Edited by Taha Selim Ustun

Published in London, United Kingdom

IntechOpen

Supporting open minds since 2005

Advanced Communication and Control Methods for Future Smartgrids
http://dx.doi.org/10.5772/intechopen.81307
Edited by Taha Selim Ustun

Contributors
Ankur Singh Rana, Mohd Asim Aftab, S.M. Suhail Hussain, Ikbal Ali, Bogdan Constantin Constantin Neagu, Gheorghe Grigoraş, Ivanov Ovidiu, Daniele Tarchi, Vahid Kouhdaragh, Alessandro Vanelli Coralli, Ujjwal Datta, Akhtar Kalam, Juan Shi, David Ribo-Perez, Carlos Álvarez, Javier Rodriguez Garcia, Manuel Alcázar-Ortega, Slavisa Aleksic, Vedad Mujan, Juan Ignacio Guerrero Alonso, Enrique Personal, Antonio Parejo, Sebastián García, Antonio García, Carlos Leon, Diego Morales, Javier Bernardo Cabrera Mejia, Manuel Fernández Veiga, Ricardo Medina

First published in London, United Kingdom, 2019 by IntechOpen
IntechOpen is the global imprint of INTECHOPEN LIMITED, registered in England and Wales, registration number: 11086078, 7th floor, 10 Lower Thames Street, London, EC3R 6AF, United Kingdom
Printed in Croatia

British Library Cataloguing-in-Publication Data
A catalogue record for this book is available from the British Library

Additional hard and PDF copies can be obtained from orders@intechopen.com

Advanced Communication and Control Methods for Future Smartgrids
Edited by Taha Selim Ustun
p. cm.
Print ISBN 978-1-78984-105-3
Online ISBN 978-1-78984-106-0
eBook (PDF) ISBN 978-1-78923-818-1

An electronic version of this book is freely available, thanks to the support of libraries working with Knowledge Unlatched. KU is a collaborative initiative designed to make high quality books Open Access for the public good. More information about the initiative and links to the Open Access version can be found at www.knowledgeunlatched.org

For EU product safety concerns:
IN TECH d.o.o., Prolaz Marije Krucifikse Kozulić 3, 51000 Rijeka, Croatia, info@intechopen.com or visit our website at intechopen.com.

Meet the editor

Taha Selim Ustun received his Ph.D. degree in electrical engineering from Victoria University, Melbourne, VIC, Australia. He has been an Assistant Professor of electrical engineering with ECE, Carnegie Mellon University, Pittsburgh, PA, USA. He is currently a Researcher with the Fukushima Renewable Energy Institute, AIST (FREA), where he leads the Smart Grid Cybersecurity Laboratory. He has edited several books and special issues with international publishing houses. His current research interests include power system protection, communication in power networks, distributed generation, microgrids, electric vehicle integration, and cybersecurity in smartgrids. He has delivered several invited talks for different conferences and organizations such as the Qatar Foundation, the World Energy Council, the Waterloo Global Science Initiative, and the European Union Energy Initiative (EUEI).

Contents

Preface

Power systems are experiencing substantial changes that will alter their operation forever. Proliferation of distributed generation and the increased ability to monitor different parts of the electrical grid offer unprecedented opportunities for consumers and grid operators. Energy can be generated near the consumption point, which decreases transmission burdens and novel control schemes can be utilized to operate the grid closer to its limits. In other words, the same infrastructure can be used at higher capacities thanks to increased efficiency. Also, new players are integrated into this grid such as smart meters with local control capabilities, electric vehicles that can act as mobile storage devices, and smart inverters that can provide auxiliary support. To achieve stable and safe operation, it is necessary to observe and coordinate all of these components in the smartgrid.

All of this requires extensive communication to collect data from all corners of the network, monitor the system status, and send the necessary instructions, when needed. Considering that there are countless different devices from many manufacturers, achieving this depends on establishing a standard communication approach. Furthermore, connecting different devices that require continuous communication with different bandwidths and security levels is not easy. Novel communication topologies and optimization approaches are needed.

An important piece of the puzzle is controlling smartgrid devices for specific purposes such as Demand Side Management or Electric Vehicle Charging Coordination. Collection of data and successful monitoring of the grid will only become meaningful if there are proper solutions implemented to dispatch storage devices, coordinate EV charging, or trigger protection schemes.

This book aims to cover new approaches developed for communication and control in smartgrids. Traditional power systems use very little communication and dynamic control therefore, such solutions are direly needed to successfully achieve power system revolution.

Taha Selim Ustun (PhD)
Fukushima Renewable Energy Institute,
AIST (FREA),
Japan

Section 1

Power System Control

Chapter 1

The Optimal Operation of Active Distribution Networks with Smart Systems

Bogdan Constantin Neagu, Gheorghe Grigoraş and Ovidiu Ivanov

Abstract

The majority of the existing electricity distribution systems are one-way networks, without self-healing, monitoring and diagnostic capabilities, which are essential to meet demand growth and the new security challenges facing us today. Given the significant growth and penetration of renewable sources and other forms of distributed generation, these networks became "active," with an increased pressure to cope with new system stability (voltage, transient and dynamic), power quality and network-operational challenges. For a better supervising and control of these active distribution networks, the emergence of Smart Metering (SM) systems can be considered a quiet revolution that is already underway in many countries around the world. With the aid of SM systems, distribution network operators can get accurate online information regarding electricity consumption and generation from renewable sources, which allows them to take the required technical measures to operate with higher energy efficiency and to establish a better investments plan. In this chapter, a special attention is given to the management of databases built with the help of information provided by Smart Meters from consumers and producers and used to optimize the operation of active distribution networks.

Keywords: smart metering, active distribution networks, optimal operation, load balancing, demand response, voltage control

1. Introduction

At present, at European level, distribution networks have a high degree of automation of distribution, using industrial standards, so transition from the current situation to the active distribution networks is technically feasible. The concepts of active distribution networks (ADN) defined both in the industrial and academic environments take different forms by focusing attention on several particular issues of concern: active consumers, distributed generation, active participation in the electricity market, etc. Each of these development directions is designed to respond to a part of issues regarding the ADN, similar to the pieces of a puzzle game. It is obvious that the ultimate success of any initiative, which refers at the transition to the ADN, is determined by the presence of the smart entity that consistently places the pieces of the game in a consistent and consistent manner [1].

It is important to address the general architecture of a control system to implement and integrate new solutions in the ADN (**Figure 1**).

To facilitate the transmission of information between new smart systems and actual distribution management systems, an integrative middleware system should be devised. The flexibility of the ADN and smart monitoring and control components is still a very important issue to be addressed. By using open standards, the ADN is designed to be expanded with virtually any future functionality [1]. Data provided by the smart meters allows detailed analyses on the operation of networks, giving a strategic advantage to distribution system operators (DSOs) in identifying the network zones or distributions which have a performance below acceptable quality, maximizing the impact of profitable investments (such as maintenance works, investments in new equipment and innovative technologies, replacing sub- or over-sized distribution transformers from the MV/LV electric substations). Also, it should be noted that these smart meters can allowed the protection of electric installations from the consumers at overvoltages, reducing the problems in case of possible incidents in the electricity grid. A meter that actively communicates with a central system can provide the important information about the position, type and magnitude of possible incidents from the network, reducing the time for intervention staff and discomfort for customers as some interventions can be made remotely [2]. The smart meters are integrated into a computerized application (smart metering system) so they can be managed centrally and remotely (**Figure 2**). In the ADN the benefits are win-win between the actors (DSO, consumers and energy producers from the renewable sources integrated into the network).

The issues such as the real-time update of consumer data on smart grids, or the integration of energy storage solutions (a critical issue in the case of discontinuous renewable energy) could be addressed by DSOs. It is estimated that ADN, summing up and extrapolating the individualized flexibility of smart meters, will be more versatile in monitoring power flows and adapting dynamically to energy consumption, helping the load balancing on the phases. The bidirectional communication is

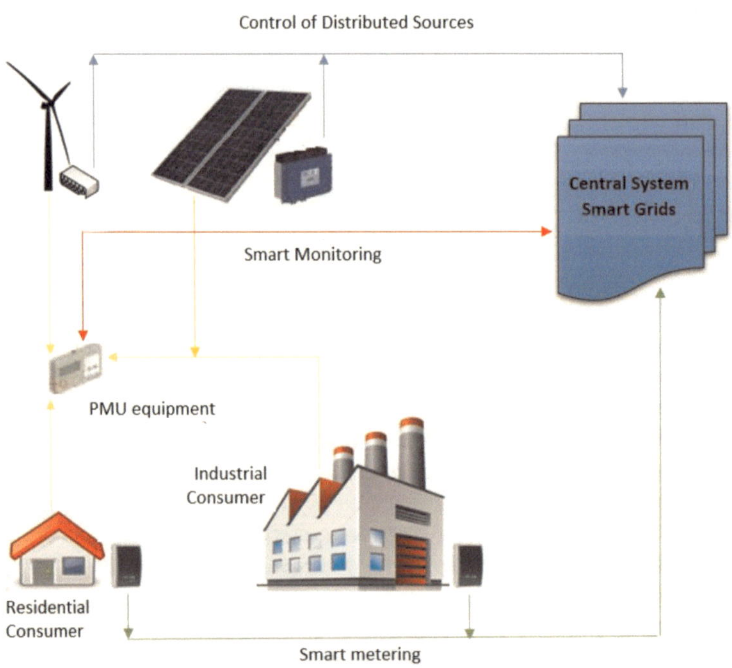

Figure 1.
The general architecture of a control system in active distribution networks [1].

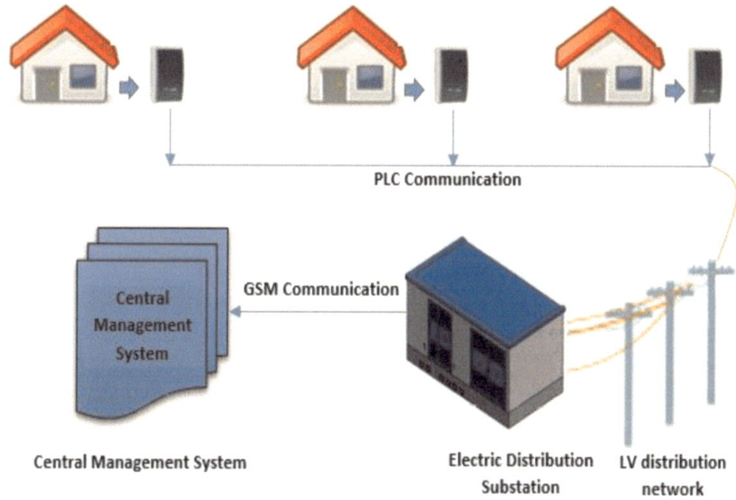

Figure 2.
The communication between the smart metering and management systems.

possible between central system from the DSO and smart meters. Also, the growing ability to integrate "green" generating unit into the network could be complemented with meteorological forecasting functions, and estimations regarding the variation in photovoltaic and wind energies could be correlated, at central level, with the daily forecasting of consumption or distributed energy (correlating with market trends through day-ahead market indicators) [3].

The current shift from fossil/nuclear to large-scale renewable energy sources (RES) brings new challenges in grid operation. The unpredictability of wind farm generation must be alleviated by DSOs with a higher flexibility of traditional generation sources and improved congestion management algorithms [4]. Also, with the increasing penetration of small distributed energy generation sources in the residential sector, the traditional consumers become prosumers, entities who generate electricity locally for their own use, and want to sell the excess power on the market [5]. For enabling the access of prosumers in the market, regulators, DSOs need to work together to create the technical infrastructure, trading regulations and management procedures for Distributed Generation (DG) sources and Demand Side Management (DSM) [6]. Inside the DSM paradigm, Demand Response (DR) is a tool that can be used by DSOs for improving system security and supply quality when operating at peak load or under restrictions imposed by the presence of RES. DR focuses on load reduction for short time intervals (e.g., hours) at consumer sites, by voluntary or automated disconnection of significant loads. To engage in DR programs, consumers or prosumers need to be equipped with Smart Metering infrastructures and Energy Management Systems (EMS), capable of automatically managing the demand and generation at household or microgrid level.

DR initiatives are currently applied for industrial consumers, which can reschedule their technological processes by shifting the operation of high-demand loads away from peak load hours. In the residential sector, DR implementation is in an incipient stage, due to consumer unawareness or lack of interest, high cost of infrastructure at the consumer side or lack of regulations or market framework [7]. One key factor for enabling the development of residential DR is the emergence of aggregators, local DSOs or independent players, which can cumulate the load reduction from several small consumers or prosumers and manage entire LV/MV network areas for DR as single entities [8]. For this purpose, aggregators can use optimization algorithms which distribute the load disconnected because of DR in a way that the technical parameters of the distribution network, such as active power losses, phase loading or bus voltage level, are kept in acceptable intervals or improved.

Voltage level control is an essential process in secure and efficient active distribution network (ADN) operation [9]. The ADN were built one century ago and they have been renewed for decades to respond to changes of end-user needs. The electricity is produced in classical grids by the central power plants, transmitted and delivered through ADN to the end-user in a one-way direction [10]. LV ADN s supply a large number of one-phase consumers, connected in a three-phase grid. Because the number of consumers and their load behavior presents a continuously dynamic, the load pattern of the three phases of the grid is different. One of the cheapest measures that a DSO can take is to optimize the steady state through voltage control and power losses and voltage drop minimization. Thereby, the real operation state of an ADN is unbalanced, and in this type of grid, the voltage control represents a relevant index, especially for LV grids, which are frequently built using OHLs mounted on poles, with supply paths extending more than 1–2 km in length. The remainder of this chapter is organized as follows. Section 2 treats the phase load balancing problem in ADN. Section 3 presents a new approach for Demand Response in ADN, and Section 4 proposes a simple method for voltage control in the real AND. For all proposed approaches, their implementation and the obtained results are discussed.

2. Phase load balancing in active distribution networks

2.1 Smart devices in phase load balancing

In the active distribution networks to operate in balancing symmetric regime, the currents on the three phases should have equal values. But, due to the unequal distribution of the consumers amongst the three phases along with variations in their individual demand appear the unequal loading of phases the so-called "current unbalance" [9]. In this context, the DSOs should take the measures by installing, besides the smart meter, a device that allows switching from phase to phase in order to balance the phases. This measure should lead at the minimization of active power losses, which represents the cheapest resource of DSOs in order to improve the energy efficiency of distribution networks [10]. In [11] is presented a constructive variant for a digital microprocessor-based device. The principle is easy, namely, for this device, a trigger module based on the minimum and maximum voltage thresholds is set so that the load to switch from the service phase to other if these thresholds are violated. The principle structure is presented in **Figure 3**.

The device is connected to the four-wire three-phase network (see **Figure 3**) through inputs 1–4 at the phases a, b, c, and the neutral (N). If it is assumed that the phase a is initial connected phase of the consumer, the voltage in this phase is monitored to be within the thresholds set. Also, the presence and voltage value of on the other two phases phase is monitored and if the voltage value on phase a fall outside the thresholds, the device will switch quickly on the phase with the higher value of voltage, but inside of thresholds (a switching delay is not more than 0.2 s) [11]. The switching process has the following succession from the phase a to b, from b to c. In [12] is presented another structure of a three-phase unbalanced automatic regulating system whose operation principle is based on the real-time monitoring and processing of three-phase current that is measured with the help of an external current transformer. A smart module equipped with a microprocessor will determine if the distribution network has a load unbalance on the three phases, then will

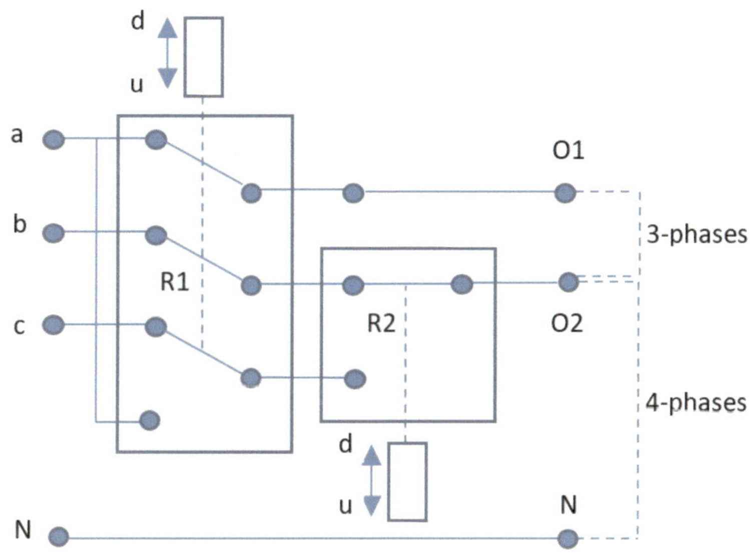

Figure 3.
The structure of digital balancing system.

determine which will be the new allocation of the consumers on phases such that the unbalance degree to be minimum. This objective can be obtained if the consumers with the higher values of the absorbed current are switched on the phase with a smaller current. At the consumers, the switch unit has in its structure a thyristor and magnetic latching relay. The role of thyristor is cut off by zero switching at the moment of input and removal, and the magnetic latching relay is switched on. The main advantages of thyristor are represented by inrush characteristic and short conduction time, because they do not lead to the generation of heat. Magnetic relay has no impact on distribution network, and it is an ideal three-phase unbalance control switch. The structure of three phase unbalance automatic regulation system is presented in **Figure 4**. The data concentrator gives the commutation command at those switch units which must transfer the consumer to the current phase on a specified phase such that to ensure as low as possible unbalance degree at the level of network.

Another structure of a smart device to connect a consumer at the distribution network is presented in [13], see **Figure 5**. According to the proposed structure, the

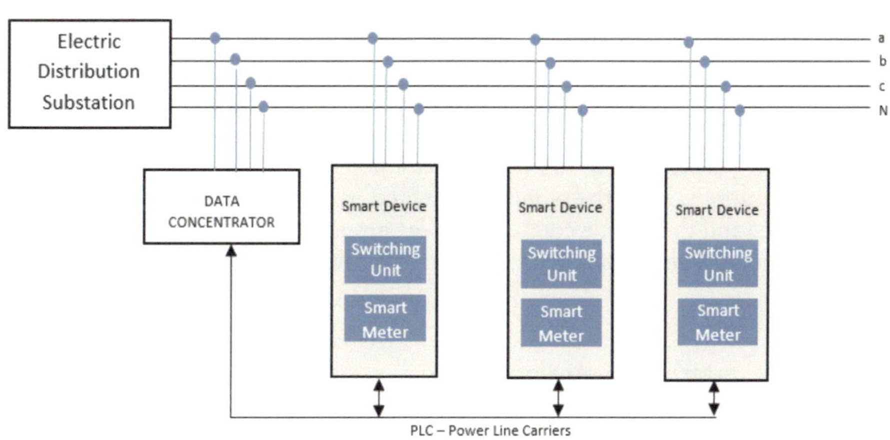

Figure 4.
The structure of smart phase microprocessor-based device.

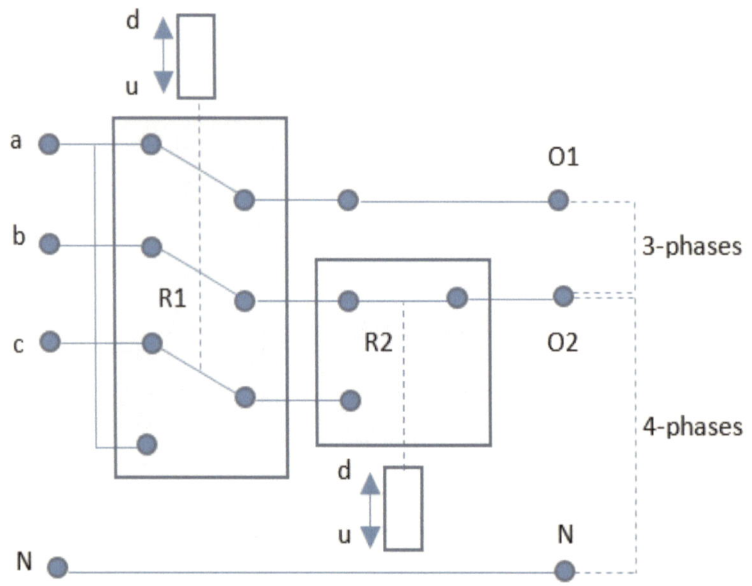

Figure 5.
The smart structure with the phase selector, [6].

Relays	R1		R2		Outputs		230 V output	
Position	d	u	d	u	O1	O2	3-phase	4-phase
1	X		X		a	b	a-b	b-N
2	X			X	a	c	a-c	c-N
3		X	X		b	c	b-c	c-N
4		X		X	b	a	b-a	a-N

Table 1.
The logic of phase selection.

smart meter is provided with a phase selector by means of which the outputs can be switched from one phase to another. In this way, when there are many 1-phase consumers connected to the distribution network, the DSO can remotely control the phase selectors in order to allocate the load over the different phases such that the unbalance degree to be minimum. In this way, a more even spreading of the load on the three phases of the distribution network can be achieved, see **Table 1** where is presented the logic of phase selection. 3-phases the output is connected to O1 and O2, respectively in the case 4-phases the output is connected to O2 and O2 The device send at the central system information about the power consumption and state (ON/OFF), which can send back the parameters for establishing the phase switching operations, after the scheme presented in **Figure 5**. Depending on the type of devices and the choice communication support, the DSOs can obtain a reliable structure, which can make the transition toward the active distribution networks.

2.2 The smart metering-based algorithm

In this paragraph, an algorithm to solve the phase load balancing (PLB) problem using a heuristic approach is proposed. This is applied to find the optimal connection phase of the 1-phase consumers such that the unbalance degree at the level of each pole to be minimum. The algorithm is based on knowing the topology of active distribution network when it will be implemented. The input data are referred at the number of poles (connection points), connected phase of each consumer, the pole when is connected the consumer, the type of consumer (1-phase or 3-phases) and load profiles provided by the smart meters. If the smart meter cannot communicate with the central unit then the algorithm will typical profiles associated to consumers without smart meters, based on the energy consumption categories and day type (weekend and working), knowing the daily energy indexes. The objective is finding the optimal phase connection for all consumers using the expression of current unbalance factor (CUF). Ideally, the value of this factor should be 1.00. But these values are very difficult to be obtained from the technical reasons and by the dynamic of loads. Thus, in most cases the obtained values will close to 1.00. The CUF factor could be evaluated using the following equation [9, 10], and the value should be under 1.10 p.u:

$$CUF_h^{(p)} = \frac{1}{3}\left[\left(\frac{I_{a,h}^{(p)}}{\overline{I}_{(a,b,c),h}^{(p)}}\right)^2 + \left(\frac{I_{b,h}^{(p)}}{\overline{I}_{(a,b,c),h}^{(p)}}\right)^2 + \left(\frac{I_{c,h}^{(p)}}{\overline{I}_{(a,b,c),h}^{(p)}}\right)^2\right] \quad p = 1, \dots, N_p, \quad h = 1, \dots, T \tag{1}$$

$$\overline{I}_{(a,b,c),h}^{(p)} = \frac{1}{3}\left(I_{a,h}^{(p)} + I_{b,h}^{(p)} + I_{c,h}^{(p)}\right) \quad p = 1, \dots, N_p, \quad h = 1, \dots, T \tag{2}$$

$$I_{a,h}^{(p)} = \sum_{k=1}^{Nc_a^{(p)}} i_{a,k,h}^{(p)}; \quad I_{b,h}^{(p)} = \sum_{k=1}^{Nc_b^{(p)}} i_{b,k,h}^{(p)}; \quad I_{c,h}^{(p)} = \sum_{k=1}^{Nc_c^{(p)}} i_{c,k,h}^{(p)} \quad p = 1, \dots, N_p, \quad h = 1, \dots, T \tag{3}$$

$$Nc^{(p)} = Nc^a{}_{(p)} + Nc^b{}_{(p)} + Nc_c^{(p)} \quad p = 1, \dots, N_p \tag{4}$$

where: a, b, and c indicate the three phases of network; $\bar{I}^{(p)}{}_{(a,b,c),h}$—the average phase current at the pole p and hour h; $I^{(p)}{}_{a,h}$, $I^{(p)}{}_{b,h}$, $I^{(p)}{}_{c,h}$—the total currents of phases a, b and c at pole p and hour h; $i^{(p)}{}_{a,k,h}$—the current of consumer k connected on the phase a, at the pole p and hour h; $i^{(p)}{}_{b,l,h}$—the current of consumer l connected on the phase b, at the pole p and hour h; $i^{(p)}{}_{c,m,h}$—the current of consumer m connected on the phase c, at the pole p and hour h; $Nc^{(p)}{}_a$, $Nc^{(p)}{}_a$, and $Nc^{(p)}{}_a$—the number of consumers connected on the phases a, b, and c, at the pole p; $Nc^{(p)}$—the total number of consumers connected at the pole p; N_p—the number of poles from the network; T—analysed period (24 h for a day or 169 h for a week).

The proposed algorithm has as start point the final poles and tries to balance the load on each phase at all poles until at the LV bus of the supply electric substation. The dynamics of unbalance process is represented by the switching from a phase on one from the other two phases (for example, from phase a to phases b or c) of some consumers such that the factor CUF to have a minimum value at the level of each pole and hour. In **Table 1**, all possible combinations in two distinct cases (3-phases and 1-phase) are presented.

Starting from the last pole N_p, depending on the initial connection of the consumers, the factor CUF could have values between 1 and 2. The minimum value, equal with 1, can be obtained in the ideal case (perfectly balanced), when the sum of phase currents corresponding the consumers are identical, and the maximum value 2 corresponds to the maximum unbalancing when only one phase current has a high value while the other two the phase currents have the values equal with 0 or close to 0. Finally, for the factor *CUF* on the LV side of the electric distribution substation (link with external grid) it is obtained the minimum value, very close to 1.0.

The minimization of the deviation between phase currents, at the level of each connection pole p (p = 1, …, N_p) at each hour h, represents the objective of the balancing problem, [7, 8]:

$$\min(\varepsilon) = \min\left(CUF_h^{(p)}\right) \quad p = 1, \ldots, N_p, \quad h = 1, \ldots, T \tag{5}$$

The problem is solved with the combinatorial optimization. Generally, a combinatorial problem is solved by total or partial enumeration of the set of its solutions (noted with Ω) [10]. In the Total Enumeration method, finding the optimal allocation $x^* \in A$, where A is the set of admissible solutions, requires the generation of all possible combinations of values given to the variables, for all elements from the set Ω, see **Table 2**. The partial enumeration approach is characterized by finding the optimal solution x^* by generating the some part from the Ω and adopting the assumption that in the remained part does not contain the optimal solutions. Regardless of the enumeration scheme, once an element $x \in \Omega$ is generated, the following two steps are performed: (1) It is investigated if element $x \in A$; if NO another element in Ω is generated. If YES, go to the next step; (2) Compare the current value of the objective function with the obtained value for the best element found in step 1; if the value of the objective function is improved (in the optimal sense), x is retained as the best item found in the set A.

Otherwise, x is dropped and a new element of Ω is generated. It is very important to highlight that the generation of the set Ω or even a part of this set does not mean the memorization of the generated elements for two reasons: there are many and then unnecessary (except the best element found in a certain iteration of the enumeration). The flow chart of the proposed algorithm is given in **Figure 6**.

To be implemented in the active distribution networks, a system with the structure presented in **Figure 4** should be used. The system contains the smart equipment installed at the consumers consisting two components and the data concentrator with an attached software infrastructure which integrate the proposed algorithm. The communication between smart equipment and data concentrator could be ensured by Power Line Carriers (PLC). From the consumers the transferred data refer at the absorbed load (current or active/reactive powers) and the connection phase. The data concentrator will transmit to each consumer the new connection phase.

2.3 Case study

The proposed method has been tested on a real distribution network from a rural area, see **Figure 7**. The main characteristics of network (poles, total length, cable type, cable section, sections length, number, type (single/three phase) and connection are indicated in **Table 3**. The connection phase of each consumer reflects the

Phases	Initial allocation	Final allocation
3-phases	[*a* | *b* | *c*]	[*c* | *a* | *b*] or [*b* | *c* | *a*] or [*a* | *b* | *c*]
1-phase	[*a* | O | O]	[O | *a* | O] or [O | O | *a*] or [*a* | O | O]
	[*b* | O | O]	[O | *b* | O] or [O | O | *b*] or [*b* | O | O]
	[*c* | O | O]	[O | *c* | O] or [O | O | *c*] or [*c* | O | O]

Table 2.
Phase switching combinations for CUF minimization.

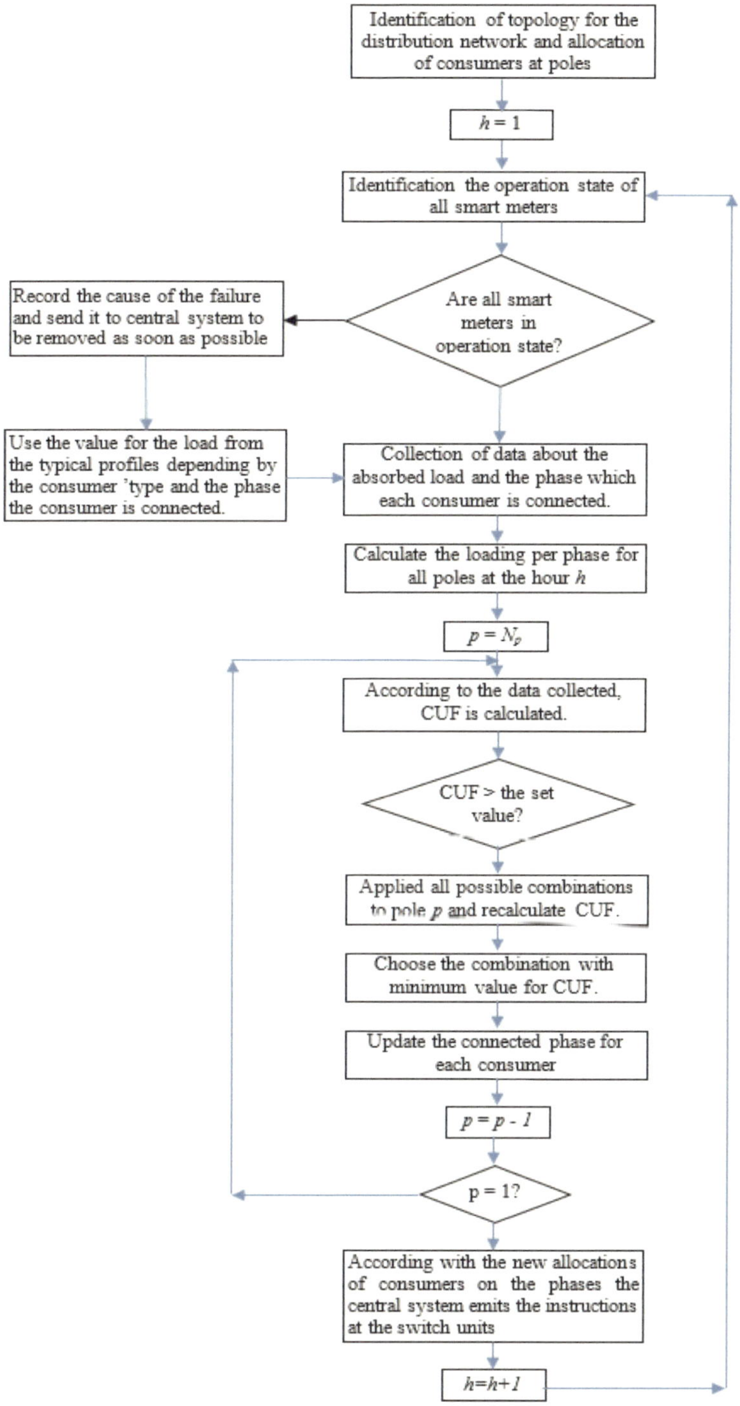

Figure 6.
The flow-chart of proposed algorithm.

situation real identified through visual inspection. The load profiles for each consumer integrated into the Smart Metering system were imported for the analysis period (27December 2017–2 January 2018). The loadings on each phase at the pole level, starting with the last pole and reaching at LV side of the electric substation, were calculated. The power flows on the three phases over the 24 h time interval on

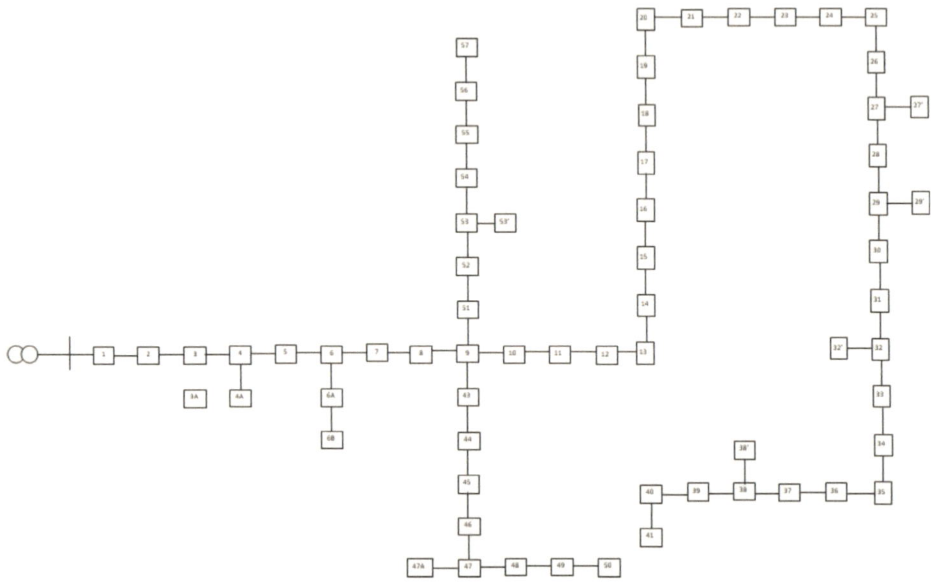

Figure 7.
The topology of test LV active distribution network.

Number of poles	Total length [m]	Data about consumers				Data about conductors		
		R	S	T	Three-phases	Type	Section [mm^2]	Length [m]
67	2560	33	28	17	6	Classical	3×35 + 35	720
		84					3×50 + 50	1840

Table 3.
The main characteristic for the analyzed feeder.

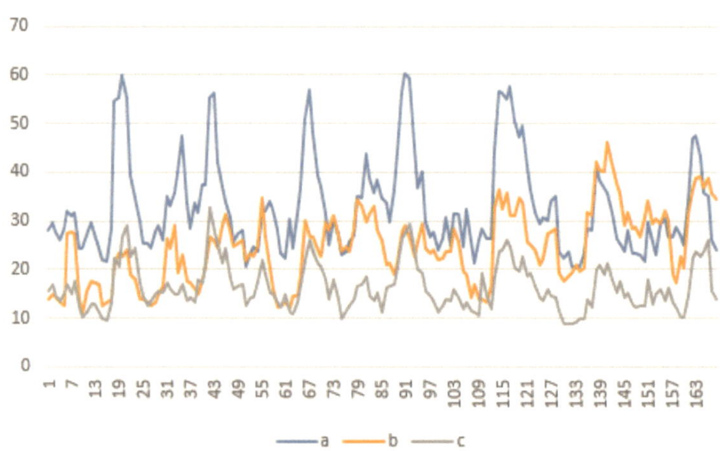

Figure 8.
The phase loading—section 0-1 [A] (initial situation—unbalanced case).

the first section are shown in **Figure 8**. It can be observed a high current unbalance degree. This degree was evaluated using the CUF factor calculated with Eq. (1).

The average value of CUF in the unbalancing case is 1.12, above the maximum admissible value (1.10). Using the proposed method, the obtained currents had the

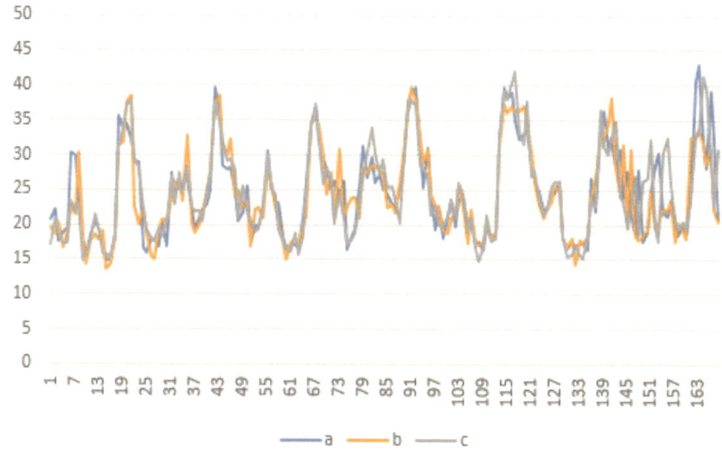

Figure 9.
The phase loading—section 0-1 [A] (final situation—unbalanced case).

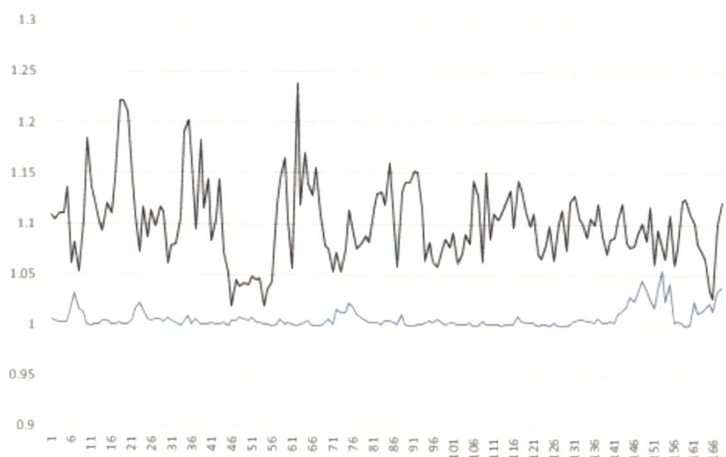

Figure 10.
Variation of CUF factor, pole no. 1.

very close values were obtained on the three phases, as can be seen in **Figure 9**, and the CUF factor was reduced to the value by 1.007. The variation of the CUF factor in the analyzed period for both situations is presented in **Figure 10**. Because the phase current unbalancing leads to voltage unbalancing, **Figures 11** and **12** show the phase voltage variation at the pole level in the study period. These values were obtained from the steady state calculation for each hour, in both situations (unbalanced and balanced) (**Figure 13**).

It can be observed that in the unbalanced case the minimum value of voltage is recorded at the pole no. 41, identified by the red color in the scheme, on the phase b ($U_b(41)$ = 221.8 V). Following the application of the balancing algorithm, the values of voltage on the phases of the network is approximately equal, and at the pole no. 41 41, on phase b, it was recorded an improved value $U_b(41)$ = 227.4 V, very close to the rated value (230 V). Also, the energy losses were reduced from 92.70 to 68.38kWh (by 26.23%), **Table 4**. A comparison between the energy losses on the phase and the neutral conductor in the both cases (unbalanced and balanced) is

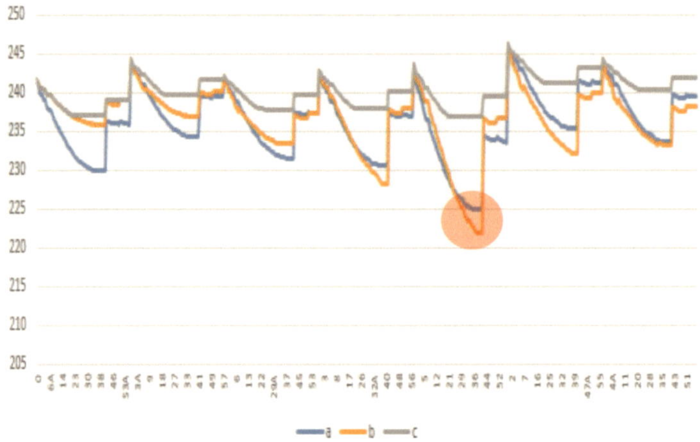

Figure 11.
Voltage variation in the nodes [V] (initial situation—unbalanced case).

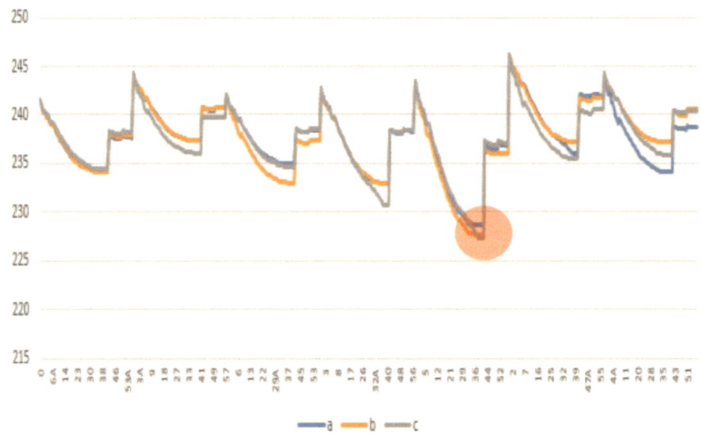

Figure 12.
Voltage variation in the nodes [V] (final situation—balanced case).

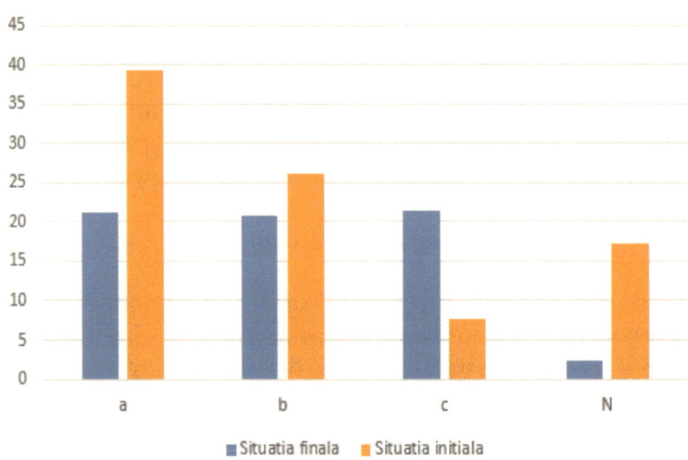

Figure 13.
Total energy losses [kWh] (unbalanced case vs. unbalanced case).

Day	Unbalanced case	Balanced case	δΔWT [kWh]
	ΔWT [kWh]	ΔWT [kWh]	
MI	10.10	7.18	2.92
JO	12.93	9.71	3.22
VI	11.29	8.49	2.8
SA	15.27	11.19	4.08
DU	16.57	11.88	4.69
LU	14.56	10.32	4.24
MA	11.99	9.59	2.4
Total	**92.70**	**68.37**	**24.33**

Table 4.
Energy losses during the analyzed period.

presented in **Figure 12**. A substantial reduction of the energy losses was obtained on the neutral conductor (approximately 86.21%, from 17.26 to 2.38 kWh).

3. Demand response in active distribution networks

While all Demand response programs encourage consumer demand flexibility by shifting or reducing load in critical time intervals, for lowering market prices and improving operation conditions in electricity transmission and distribution networks, there are several ways to achieve this goal. The literature distinguishes two main types of DR: controllable (incentive-based) and price-based [14].

The former are most restrictive DR approaches and they frequently involve direct or indirect load control, according to the curtailment level required by the coordinating entity of the program (usually, the DSO or an aggregator). Direct Load Control (DLC) is remotely enforced by the coordinating entity, a task that requires bidirectional real-time communication with the consumer site. On the other hand, price-based DR relies on consumer response to electricity price variations.

3.1 Demand response management algorithm for ADN

The involvement of residential consumers in DR programs is currently in its incipient stage. Several problems contribute to this situation. The first are the demand level of individual consumers and the need for aggregators. Residential consumers have much lower demand, compared to other consumer categories, such as industry. In rural underdeveloped areas, most consumers achieve less than 1 kW power draw. Because electricity markets require minimum demand reduction biddings of 100 kW and more [8], the participation of residential consumers to DR programs is feasible only to households with higher demand, managed by aggregators who can achieve the minimum DR levels required by the market.

Another key factor is the user comfort. As a general rule, residential consumers are not willing to sacrifice to a great extent their personal comfort in order to better contribute to DR. As such, a household will try to set and accomplish a DR target with minimum effort, while maintaining its comfort requirements (i.e., room air temperature). The process of dynamically optimizing appliance schedule while accounting for pre-set comfort levels, market price variation and DR signals, requires automated algorithms, known as Smart Home Energy Management

Systems (SHEMS) [15]. While the effect of the rebound load on operating frequency is negligible [16], it can be higher regarding network losses and quality of supply.

As described in [17], artificial intelligence algorithms are widely used for managing DR at LV network level. This paragraph describes a DR Management Algorithm for aggregators based on the Particle Swarm Optimization (DRMA), which investigates the effect of rebound load in a LV distribution network, taking into account consumer demand levels, comfort and privacy preferences. The algorithm requires as input the following information:

- network data (topology, length of feeder sections, wire type, consumer phase and pole connection) and load data, given as consumer active and reactive power load profiles with a known (e.g., hourly) sampling: $P_{i,h}$, $Q_{i,h}$, $h = 1...24$;

- the *DR* interval set by the aggregator: $Int_{DR} = [h1, h2], h1, h2 \in [1..24], h1 < h2$;

- the DR signal magnitude for the entire network, *targetDR*, given in percent from the network demand in each hour $h1...h2$;

- the maximum percent reduction from each consumer load i, $maxDR_i$, the same for all DR intervals, a value which consider the consumer comfort preferences, given as the maximum load that can be disconnected upon request. At each time interval h, the actual reduction demand of the aggregator must not exceed the maximum limit for any consumer i ($lDR_{i,h} \leq maxDR_i$, $h \in Int_{DR}$, $i = 1..Nc$);

- the load rebound rate RB_h, which describes the load amount which will be switched back on by the consumers after DR, at hour $h + 1$, given in percent from the reduction at hour h. For modeling consumer behavior uncertainty, the actual value of the rebound can be set randomly in an interval from $[0, RB_h]$.

Based on the consumer load and rebound data, the DRMA determines which consumers are eligible for DR load curtailment, according to their hourly demand. Only consumers exceeding a given load threshold ($P_{i,h} \geq P_{max,h}$, $h \;\; Int_{DR}\, i = 1..Nc$) will be considered for DR. A higher threshold will result into a smaller number of consumers being affected. The scheduling of appliances is performed by each consumer, using its SHEMS and comfort preferences. For privacy reasons, the aggregator receives only the load reduction $lDR_{i,h}$,. For determining the optimal DR signal for each household, the Particle Swarm Optimization algorithm [10] is used. The solutions are encoded as number vectors in which each element describes the *DR* reduction applied to the load of each eligible consumer at hour h, as described in **Figure 14**. Here, n is the number of consumers eligible for DR at hour h, and value 3 for consumer n depicts a 30% DR load reduction.

The fitness function based on which the solutions are evaluated has two factors: minimum active power losses in the LV network, and minimum difference between the expected and obtained load reduction by DR, computed for the hour h for which the PSO algorithm is running. Both are expressed in percent.

Cons. i	*Cons. j*	...	*Cons. n*
2	4		3

Figure 14.
Solution encoding for the PSO algorithm.

$$\min(\Delta P_h + M_{DR,h}), \quad h \in Int_{DR} \tag{6}$$

By this approach, it is expected that the algorithm will search for solutions where $M_{DR,h}$ is close to 0, choosing between them as optimal solution the one with the minimal value of ΔP_h. The active losses in the network ΔP_h are computed using the graph theory, with a procedure consisting of several steps. In the initialization stage of the algorithm, the branch-node connectivity matrix $[A]$ and the reference bus loads for each hour h from the DR interval are computed, using the topology of the network and the consumer data. Next, the real bus loads are determined, by subtracting the loss reduction imposed by the DR signal:

$$P_{bs,x} = P_{ref,bs,x} - P_{\mathrm{DR},bs,x}, \quad bs = 1..n \tag{7}$$

The bus active power loads are converted into bus current injections, using the nominal voltage of the network and the power factor:

$$I_{bs,x} = P_{bs,x}/(U_n \cdot \cos(\varphi_{bs})), \quad bs = 1..Nb \tag{8}$$

The branch current flows on each feeder section (branch) br and phase x are then computed, using matrix A and the bus current vector $[I_{bs,x}]$:

$$[I_{br,x}] = -inv(A) \cdot [I_{bs,x}] \tag{9}$$

The power losses in kW on each section br follow:

$$\Delta P_{br,x} = R_{a,br} \cdot I_{br}^2 \cdot K_{br} \cdot 10^{-3} \tag{10}$$

The K_{br} coefficient is used to account for the losses caused by the current flow on the neutral wire. According to Romanian standards, K_{br} is computed for each branch using the CUF factor (1), considering the phase load variation in each hour, with:

$$K_{br,h} = CUF_{br,h} \cdot \left(1 + 1.5 \cdot \frac{R_{n,br}}{R_{a,br}}\right) - 1.5 \cdot \frac{R_{n,br}}{R_{a,br}} \tag{11}$$

In Eqs. (6)–(11), $P_{bs,x}$—the aggregate active power draw in bus bs and phase x (x is a, b, c or abc), during DR; $P_{ref,bs,x}$—the aggregate reference active power draw in bus bs and phase x without DR; $P_{DR,bs,x}$—the aggregate active power reduction in bus bs and phase x, during DR; $I_{bs,x}$—the aggregate current draw in bus bs, on phase x, during DR; U_n—the nominal voltage of the network, $\cos(\varphi_{bs})$—the power factor determined from the aggregate individual active and reactive loads at bus bs; Nb—the number of buses in the network; $R_{a,br}$, $R_{n,br}$—the resistance on section br, on the active and neutral wire respectively. Powers are given in [kW], currents in [A], and voltages in [kV]. The percent active losses in the network for hour h are obtained by summing all branch losses obtained with Eq. (10) and dividing the result by the power infeed of the network:

$$\Delta P = \left(\sum_x \Delta P_{br} \Big/ \left(\sum_x P_{bs} + \sum_x \Delta P_{br}\right)\right) \cdot 100 \tag{12}$$

On the other hand, he difference between the expected and obtained load reduction by DR, $M_{DR,h}$, uses the sum of all differences between the aggregate bus power draws, in absence of and during DR (Eq. (13)).

$$M_{DR} = \left(1 - \sum_{x,\,IntDR} P_{bs,x} \Big/ \sum_{x,\,IntDR} P_{ref,\,bs,\,x}\right) \cdot 100 \qquad (13)$$

In Eqs. (7)–(13), the hourly index h was omitted for simplicity. The solution which minimizes Eq. (6) is the optimal *DR* dispatch among the eligible consumers for hour h. The PSO algorithm is executed for each hour from the *DR* interval, with the bus loads updated to account the hourly demand and the rebound after the *DR* signal from the previous hour has ended. The block diagram of the DRMA algorithm is given in **Figure 15**.

3.2 Case study

The DRMA was tested on a real Romanian distribution network from a rural area, namely network T2, from **Figure 16**. The main characteristics of the network (number of poles or buses; cable type and cross-section; feeder section lengths; number; type (single phase/three phase) and connection phase of consumers) are indicated in **Table 5**. The load profiles for all the consumers are provided by a Smart Metering system. The hourly load profile of the entire network, on each of the three phases, is given in **Figure 17**.

Analyzing the load profile of the network from **Figure 17**, the Int_{DR} interval was set as $h1 = 18$, $h2 = 22$, for a *DR* interval of 5 h, for the evening peak load time. In **Tables 6–8** are given the results corresponding to the following scenarios:

- Ref—the reference case, where no *DR* request is enforced in the network, and the demand varies according to **Figure 17**;
- DR00—*DR* with 0% rebound, the ideal case, where the load disconnected because of the DR signal is not switched back on later;
- DR50—*DR* with minimum 20% and maximum 50% rebound, when maximum half of the load disconnected in hour h will return at the network buses in $h + 1$.

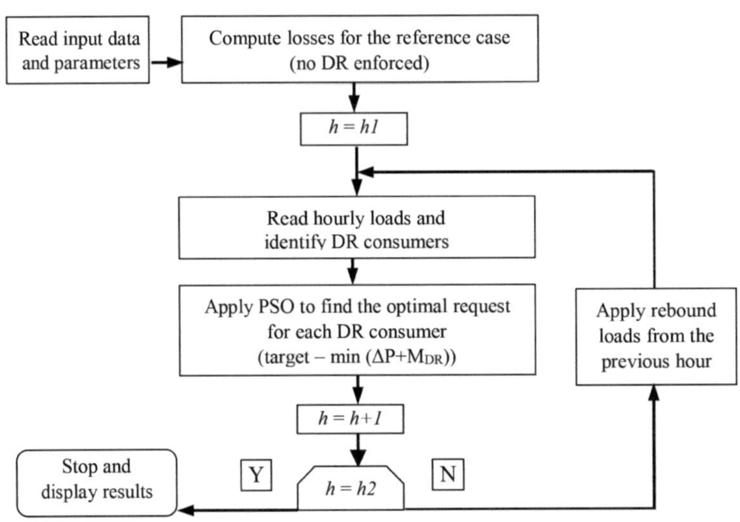

Figure 15.
The block diagram of the DRMA.

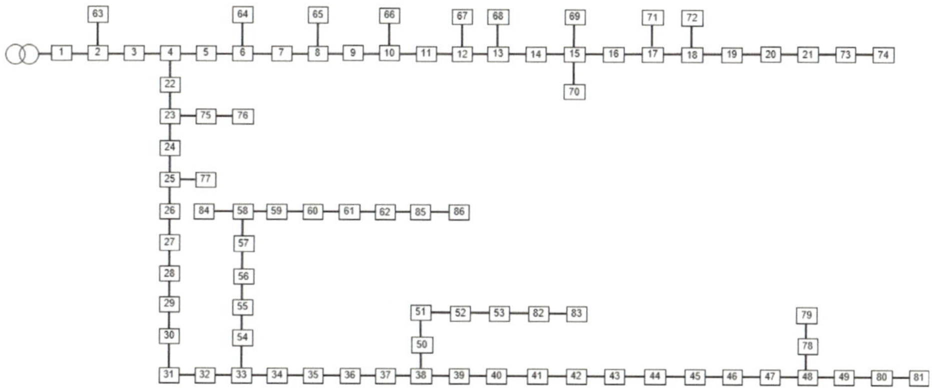

Figure 16.
The topology of the test network T2.

Number of poles	Total length [m]	Connection phase				Wire data		
		a	b	c	abc	Type	Section [mm^2]	Length [m]
86	3440	20	21	19	0	OHL Ol-Al	3 × 50 + 35	3440
		60						

Table 5.
Input data for test network T2.

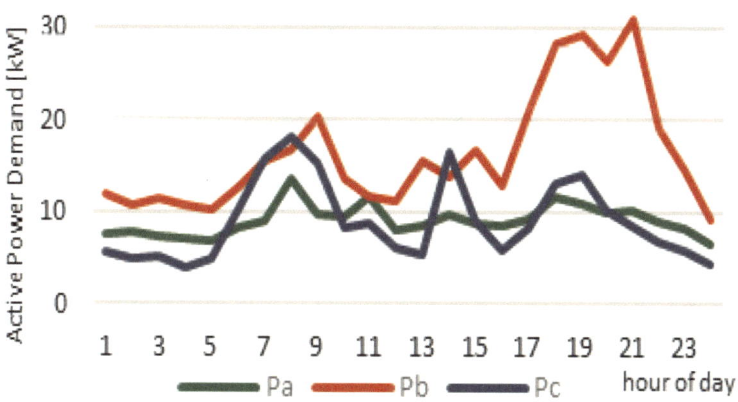

Figure 17.
Phase load in the network, on the three phases.

h	P_a [kW]	P_b [kW]	P_c [kW]	ΔP_{abc} [kW]	ΔP_a kW]	ΔP_b [kW]	ΔP_c [kW]	ΔP_{abc} [%]	ΔP_a [%]	ΔP_b [%]	ΔP_c [%]
18	11.596	28.215	13.028	**3.486**	0.820	1.869	0.797	**6.19**	6.60	6.21	5.76
19	10.981	29.093	14.132	**4.079**	0.810	2.366	0.902	**7.00**	6.87	7.52	6.00
20	10.087	26.366	10.229	**4.029**	0.761	2.735	0.533	**7.95**	7.01	9.40	4.96
21	10.285	30.877	8.494	**7.211**	1.043	5.695	0.474	**12.68**	9.21	15.57	5.28
22	9.050	19.016	6.806	**2.709**	0.789	1.704	0.216	**7.21**	8.02	8.22	3.08

Table 6.
Results for the reference case without DR (scenario ref).

h	P_a [kW]	P_b [kW]	P_c [kW]	ΔP_{abc} kW]	ΔP_a [kW]	ΔP_b [kW]	ΔP_c [kW]	ΔP_{abc} [%]	ΔP_a [%]	ΔP_b [%]	ΔP_c [%]
18	9.256	24.791	10.860	**2.249**	0.446	1.313	0.490	**4.77**	4.59	5.03	4.32
19	9.529	23.739	12.821	**2.375**	0.528	1.211	0.636	**4.90**	5.25	4.85	4.73
20	9.339	20.598	9.750	**2.066**	0.496	1.167	0.403	**4.95**	5.04	5.36	3.97
21	9.639	24.074	8.494	**3.355**	0.697	2.287	0.371	**7.36**	6.74	8.68	4.18
22	7.451	15.382	6.806	**1.441**	0.412	0.838	0.191	**4.64**	5.24	5.17	2.73

Table 7.
Results for the case DR with no rebound (scenario DR00).

h	P_a [kW]	P_b [kW]	P_c [kW]	ΔP_{abc} [kW]	ΔP_a [kW]	ΔP_b [kW]	ΔP_c [kW]	ΔP_{abc} [%]	ΔP_a [%]	ΔP_b [%]	ΔP_c [%]
18	9.499	24.791	10.620	**2.234**	0.467	1.309	0.457	**4.74**	4.69	5.02	4.12
19	10.326	24.781	13.436	**2.699**	0.601	1.372	0.726	**5.27**	5.50	5.24	5.13
20	9.585	21.456	10.430	**2.374**	0.588	1.296	0.490	**5.41**	5.78	5.70	4.48
21	10.241	25.070	8.425	**3.830**	0.853	2.607	0.370	**8.05**	7.69	9.42	4.21
22	8.255	16.265	6.835	**1.773**	0.543	1.029	0.202	**5.35**	6.17	5.95	2.87

Table 8.
Results for the case DR with 20–50% rebound (scenario DR50).

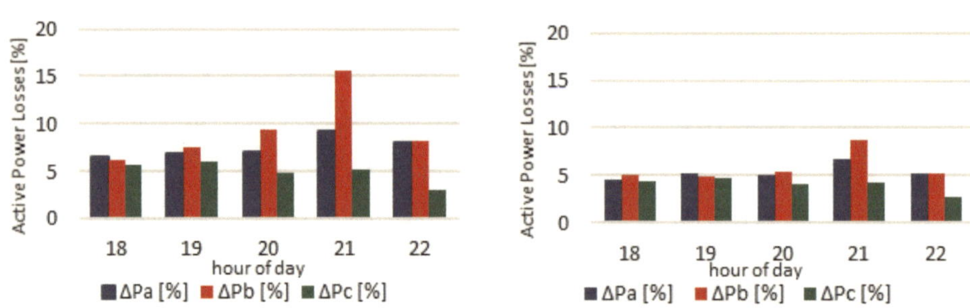

Figure 18.
Phase active power losses in scenario Ref (left), DR00 (right).

The minimum hourly load for DR-eligible consumers was set at 0.8 kW. This setting resulted in 13–23 consumers affected by the *DR* signal in the DR50 scenario, the number varying in each hour according to the bus and rebound loads. **Figure 17** shows that the phase loads are highly unbalanced, which results in higher power losses in the network, due to the excessive loading of phase *b* and the neutral wire current flow. By optimally dispatching the *DR* signal across the network in each hour, the DRMA algorithm is expected to also reduce the phase load unbalance from the reference case.

The results show that the active power losses decrease in the *DR* scenarios, more if there is no rebound load in the 19:00–22:00 interval. The best effect can be seen at the peak hour 21:00, when the losses drop from 12.68% in the reference case to 7.36% in the DR00 scenario and 8.05% in the DR50 scenario respectively. The phase loss distribution in the three scenarios, depicted in **Figure 18**, shows that a secondary effect of *DR* an improved balancing of the phase loss values. The load variation for scenario DR50 is similar to case (b).

4. Voltage control in active distribution networks

4.1 Problem statement

The voltage control strategies are sometimes a key performance indicator in ADN. In the literature, this problem is solved using pseudo-measurements. Due to the intermittent and unpredictable behavior of consumptions and distributed energy sources, the generation excess could lead to a reversed power flow, from the consumers to the supply external point [18, 19]. This drawback requires a real-time effective voltage control strategy [20], particularly under islanded operation modes, to obtain the best solutions, with reliable effects on the minimization of energy losses, and energy efficiency improvement [21, 22]. Our proposed approach uses Smart Metering information (active and reactive daily load curves).

The objective of the optimization procedure is to assess the influence of renewable sources (i.e., wind turbines) into an ADN in order to improve the voltage at the end-users and to minimize the active power losses, considering the technical constraints. The proposed approach was formulated as:

$$MinF([U],[\Psi]) = \min\left(\Delta U^{h}_{ADN}\right) + \min\left(\Delta P^{h}_{ADN}\right) \tag{14}$$

where: *minF*—the goal function;
$[U]$—the voltage magnitude vector;
$[\Psi]$—the transformers tap changing matrix;
ΔU—the voltage drops;
h = 1...T, the hourly measurement interval for the steady state;
ΔP—the active power losses.

The equality constraints coincide with the bus power balance in the ADN. For a given bus k = 1,…, N, a time sample h, and an operating states j, the equations are:

$$P^{3ph}_{h,k,j} + jQ^{3ph}_{h,k,j} = \overline{U}_{h,k,j} \cdot \overline{I}^{*}_{h,k,j} \tag{15}$$

where the active and reactive power are a sum of the three phases of the ADN:

$$P^{3ph}_{h,k,j} = P^{a}_{h,k,j} + P^{b}_{h,k,j} + P^{c}_{h,k,j};\ Q^{3ph}_{h,k,j} = Q^{a}_{h,k,j} + Q^{b}_{h,k,j} + Q^{c}_{h,k,j} \tag{16}$$

The mathematical model has the following inequality constraints:

1. Voltage allowable limits:

$$U^{\min}_{h,k,j} \le U_{h,k,j} \le U^{\max}_{h,k,j} \tag{17}$$

2. Thermal limits of the branch loadings:

$$S_{h,k,j} \le S^{\max}_{h,k,j} \ \ or\ \ I_{h,k,j} \le I^{\max}_{h,k,j} \tag{18}$$

3. The allowable reactive power of DG sources must be constrained as:

$$Q^{wind}_{h,\min} \le Q^{wind}_{h} \le Q^{wind}_{h,\max} \tag{19}$$

4. The constraints for the transformer tap changer must be in accordance with the proposed strategy, and are the following:

$$\Psi^h_{\min} \le \Psi^h_{k,j} \le \Psi^h_{\max} \quad (20)$$

where U^{min}, U^{max}—the inferior/superior voltage limit; $I_{h,k,j}$—the branch current value; I^{max}—the branch ampacity value; $S_{h,k,j}$—the branch apparent power; S^{max}—

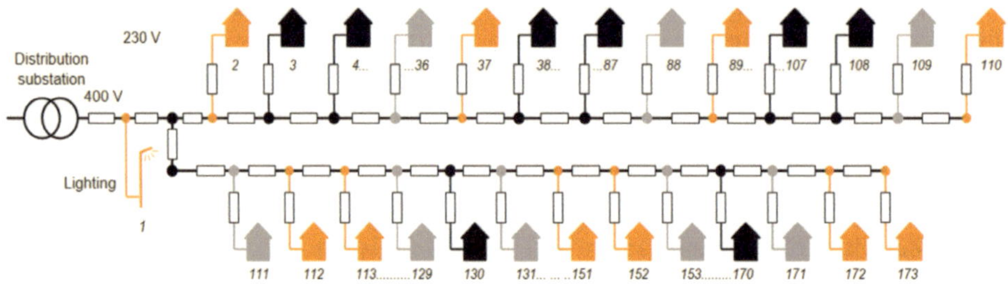

Figure 19.
Single-line diagram of the LV distribution network.

Hour/cases	Pole no. 88			Pole no. 110		
	I	II	III	I	II	III
1	0.3684	0.3924	0.3925	0.3673	0.3913	0.3915
2	0.3701	0.3928	0.3929	0.3691	0.3919	0.3919
3	0.3721	0.3933	0.3933	0.3712	0.3924	0.3925
4	0.3718	0.3932	0.3932	0.3708	0.3923	0.3923
5	0.3719	0.3933	0.3933	0.3710	0.3924	0.3924
6	0.3764	0.3947	0.3947	0.3755	0.3939	0.3939
7	0.3715	0.3936	0.3935	0.3705	0.3926	0.3925
8	0.3681	0.3927	0.3927	0.3670	0.3916	0.3916
9	0.3653	0.3918	0.3920	0.3641	0.3907	0.3909
10	0.3640	0.3913	0.3916	0.3628	0.3901	0.3905
11	0.3584	0.3896	0.3897	0.3571	0.3883	0.3884
12	0.3617	0.3902	0.3907	0.3605	0.3890	0.3896
13	0.3591	0.3894	0.3902	0.3579	0.3882	0.3891
14	0.3600	0.3900	0.3907	0.3587	0.3887	0.3895
15	0.3593	0.3897	0.3905	0.3580	0.3884	0.3893
16	0.3646	0.3914	0.3917	0.3634	0.3902	0.3906
17	0.3591	0.3897	0.3905	0.3578	0.3884	0.3892
18	0.3534	0.3883	0.3885	0.3519	0.3868	0.3869
19	0.3583	0.3900	0.3901	0.3568	0.3886	0.3887
20	0.3628	0.3914	0.3917	0.3615	0.3901	0.3904
21	0.3561	0.3906	0.3898	0.3545	0.3892	0.3882
22	0.3482	0.3895	0.3880	0.3462	0.3879	0.3861
23	0.3493	0.3880	0.3882	0.3474	0.3861	0.3863
24	0.3642	0.3917	0.3920	0.3629	0.3905	0.3908

Table 9.
Voltage magnitude for the two representative busses [kV].

Case	W_{load} [kWh]	ΔW_{loss} [kWh]	ΔW_{loss} [%]	Energy savings [kWh]
Case I (base)	442.47	78.95	15.14	521.42
Case II (DG connected)	442.47	62.36	12.35	504.83
Case III (DG + AVR)	442.47	61.13	12.13	503.60

Table 10.
Comparison between the simulation cases.

the maximum apparent power on branch; Q^{wind}—the reactive power from the DG source; Q_{min}, Q_{max}—the allowable reactive power limits; Ψ—the tap position of the transformer.

4.2 Case study

The voltage control approach proposed above was tested on a real ADN with 163 residential consumers, presented in **Figure 19**. It must be highlighted that the tested ADN already includes two connected small-scale renewable sources.

In order to demonstrate the capabilities of the proposed voltage control strategy, three scenarios for simulation using MATLAB environment were considered:

- First, the base case (Case I) without small-scale sources and AVR control (with the initial tap position).

- The second case (Case II) considers the two real wind generators (2 × 5 kW) connected into the AND.

- The last case (Case III) uses the voltage control strategy (14)–(20).

Case II is proposed for assessing the influence of the DG sources on the voltage and power losses magnitude in a real ADN. In addition, Case III follows the improvement of voltage magnitude based on a coordination between the generation of the distributed sources and the automation distribution devices.

The results regarding the voltage magnitude in the three considered cases, are given in **Table 9**, only for representative connected points of DGs: pole no. 88 and the last ADN bus, pole no. 110. The daily energy losses are presented in **Table 10**, where W_{load} is the total energy required by the consumers.

It can be observed in **Table 10** a reduction of energy losses, with over 3%, from 15.14 to 12.13% with energy savings of about 17.82 kWh for the entire ADN.

5. Conclusions

The active distribution networks will be developed based on the improved actual infrastructure with the main advantage regarding the bidirectional communication between supplier and consumers. This makes possible a supervising and control at an advanced level of the smart systems which will be integrated inside their.

The chapter aimed to highlight the advantages of introducing the smart systems in the active distribution networks that lead to an optimal operation regarding the phase load balancing, voltage control and demand response with benefits for both DSO and consumers. The offered solutions are based on the information provided by the smart meters, these being an important link between the consumers, dispatch centers of DSOs, renewable sources, and the smart systems integrated in the networks. The case studies based on the pilot active distribution networks belonging to a DSO from Romania, emphasized the importance of integrating the smart devices so that the control to make easy and the transition to self-control networks to be smooth. The obtained results allow us to expect that in a short time the expression "active" will be used for all distribution networks.

Author details

Bogdan Constantin Neagu*, Gheorghe Grigoraş and Ovidiu Ivanov
Department of Power Engineering, Gheorghe Asachi Technical University of Iasi, Romania

*Address all correspondence to: bogdan.neagu@tuiasi.ro

References

[1] Adrem Invest. Coordinating SMART GRID Initiatives and Developments with Current Challenges and Evolutions of Power Systems (in Romanian). Available from: http://slideplayer.fr/user/15806718

[2] E-distribution. Upgrade Programmes (in Romanian). Available from: https://www.e-distributie.com/ro-RO/Pagini/Programe-de-modernizare.aspx

[3] Smart Metering—Under consecration (in Romanian). Available from: https://www.ttonline.ro/revista/energie/smart-metering-in-curs-de-consacrare

[4] Variable Renewables Integration in Electricity Systems: How To Get It Right, World Energy Perspectives Renewables Integration [Internet]. 2016. Available from: https://www.worldenergy.org/news-and-media/news/variable-renewables-integration-in-electricity-systems-how-to-get-it-right/ [Accessed: 02 May 2019]

[5] Parag Y, Sovacool BK. Electricity market design for the prosumer era. Nature Energy. 2016;**1**(4):1-6. DOI: 10.1038/nenergy.2016.32

[6] Towards Smarter Grids: ENTSO-E Position Paper on Developing TSO and DSO Roles for the Benefit of Consumers [Internet]. 2015. Available from: https://www.entsoe.eu/ [Accessed: 2019-05-02]

[7] Paterakis NG, Erdinç o CJPS. An overview of demand response: Key-elements and international experience. Renewable and Sustainable Energy Reviews. 2017;**69**:871-891. DOI: 10.1016/j.rser.2016.11.167

[8] SEDC. Explicit Demand Response in Europe-Mapping the Markets [Internet]. 2017. Available from: https://www.smarten.eu/wp-content/uploads/2017/04/SEDC-Explicit-Demand-Response-in-Europe-Mapping-the-Markets-2017.pdf [Accessed: 02 May 2019]

[9] Mansani S, Udaykumay RY. An optimal phase balancing technique for unbalanced three-phase secondary distribution systems. In: 2016 IEEE 7th Power India International Conference (PIICON); 2016. pp. 1-6. DOI: 10.1109/POWERI.2016.8077172

[10] Ivanov O, Grigoraș G, Neagu BC. Smart metering based approaches to solve the load phase balancing problem in LV distribution networks. In: International Symposium on Fundamentals of Electrical Engineering (ISFEE 2018); Bucharest, Romania; in Press

[11] Li Y, Gong Y. Design of three phase load unbalance automatic regulating system for low voltage power distribution grids. In: 2018 International Conference on Smart Materials, Intelligent Manufacturing and Automation; 24–26 May 2018; Nanjing, China; 2018

[12] Novatek electro. Universal Automatic Electronic Phase Switch PEF-301. Available from: https://novatek-electro.com/en/products/phase-selector-switch/universal-automatic-electronic-phase-switch-pef-301.htm

[13] Henderieckx L. Smart Metering Device With Phase Selector. Available from: https://patentimages.storage.googleapis.com/d4/d5/f5/9ab891385ce880/US20120078428A1.pdf

[14] Asadinejad A, Tomsovic K. Optimal use of incentive and price based demand response to reduce costs and price volatility. Electric Power Systems Research;**144**:215-223. DOI: 10.1016/j.epsr.2016.12.012

[15] Shareef H, Ahmed MS, Mohamed A, Hassan EA. Review on home energy

management system considering demand responses, smart technologies, and intelligent controllers. IEEE Access. 2018;**6**:24498-24509. DOI: 10.1109/ACCESS.2018.2831917

[16] Lütolf P, Scherer M, Mégel O, Geidl M, Vrettos E. Rebound effects of demand-response management for frequency restoration. In: Proceedings of the 2018 IEEE International Energy Conference (ENERGYCON'18); 3–7 June 2018; Limassol, Cyprus; 2018. pp. 1-6. DOI: 10.1109/ENERGYCON.2018.8398849

[17] Rahman MN, Arefi A, Shafiullah GM, Hettiwatte S. A new approach to voltage management in unbalanced low voltage using demand response and OLTC considering consumer preference. International Journal of Electrical Power & Energy Systems. 2018;**99**:11-27. DOI: 10.1016/j.ijepes.2017.12.034

[18] Xu G, Yu W, Griffith D, Golmie N, Moulema P. Toward integrating distributed energy resources and storage devices in smart grid. IEEE Internet of Things Journal;**4**(1):192-204. DOI: 10.1109/JIOT.2016.2640563

[19] Li Y, Tian X, Liu C, Su Y, Li L, Zhang L, Sun Y, Li J. Study on voltage control in distribution network with renewable energy integration. In: 2017 IEEE Conference on Energy Internet and Energy System Integration (EI2); 2017. pp. 1-5. DOI: 10.1109/EI2.2017.8245755

[20] Kennedy J, Eberhart RC. Particle swarm optimization. In: Proceedings of the IEEE International. Conference on Neural Networks (ICNN'95); 27 November–1 December 1995; Perth, Australia; 1995. pp. 1942-1948

[21] Ranamuka D, Agalgaonkar AP, Muttaqi KM. Online voltage control in distribution systems with multiple voltage regulating devices. IEEE Transactions on Sustainable Energy. 2014;**5**(2):617-628. DOI: 10.1109/PESGM.2014.6939417

[22] Capitanescu F, Bilibin I, Romero Ramos E. A comprehensive centralized approach for voltage constraints management in active distribution grid. IEEE Transactions on Power Apparatus and Systems. 2013;**29**(2):933-942. DOI: 10.1109/TPWRS.2013.2287897

Chapter 2

Optimal Power Flow Solution in Smart Grid Environment Using SVC and TCSC

Ankur Singh Rana, Mohit Bajaj and Shrija Gairola

Abstract

Flexible AC transmission system devices (FACTS) are most promising controllers in present day scenario when it comes to power transmission in long distances in smart grids. FACTS devices provide system stability, midpoint voltage support and reactive power control in grid interconnections. Conventionally, power flow algorithm was used to evaluate the rating of FACTS devices by taking consideration of magnitude of voltage and phase angle as independent variables. Nowadays, FACTS device rating is evaluated with a new framework called optimal power flow. This chapter provides a comparison for optimal power flow, with or without FACTS devices such as static VAR compensator (SVC) and thyristor controlled series capacitor (TCSC), in terms of cost saving and loss reduction in smart grid scenario.

Keywords: FACTS controllers, smart grid, SVC, TCSC optimal power flow, Lagrangian function

1. Introduction

In power system, interconnections were primarily used for pooling of power between power plants and load centers along with the added advantages of reduction of overall generation capacity, minimum generating cost with increased reliability and better utilization of energy reserves.

Advancement in the FACTS technology used for renewable energy generation as well as for the transmission and distribution network interconnections ask for the upgrade should be smart enough to cope up for the solution.

Smart grid concept came into the effect with the enhancement of technologies like new and renewable energy sources, also the power transfer capabilities get increased manifold in case of both transmission and distribution system. Also the use of smart grid technology provides more flexible, stable and efficient operation [1]. In smart grid interconnection, FACTS devices will increase the power transfer capacity between existing transmission lines without erecting a new line [2]. FACTS controller either reduce impedance of the line (by injecting voltage drop) or increase the phase angle in turn increasing the active power transfer in power system. Also, FACTS device does the reactive power compensation by injecting a current to the existing system. Conventionally, controlled mechanical switches with large switching time were used to connect the compensators to the transmission line. During the time of fault, power system needs a fast recovery and to fulfill this requirement fast acting power electronics base FACTS devices are used in place of mechanical switches. Apart from this, there are some major benefits of FACTS devices listed below:

- Increase in power transfer capabilities of transmission lines.
- Provide voltage support along the line.
- Provide reactive power compensation in mid pints as well as at receiving end of the line.
- Enhance the dynamic as well as steady state stability of the system interconnections.
- Improvement in power factor.

On the basic of its placement in transmission line, FACTS controllers are classified in four categories as shown in **Figure 1**.

First is **series controllers** which inject voltage to the system. It also offers variable impedance to the system. These controllers absorb or deliver active as well as reactive power in the line. Examples of series FACTS controllers are: GTO controlled series capacitor (GCSC), thyristor controlled series capacitor (TCSC), static synchronous series compensators (SSSC), etc. Second is **shunt controllers** which inject current to the system. If the phase angle between injected current and line voltage is 90 degrees, then the device will deliver or absorb reactive power only. Static VAR compensators (SVC), static synchronous compensators (STATCOM) are the examples of this type of controller. **Combined series-series controllers** are third type of FACTS controller which are used where more than one transmission lines needs active and reactive power compensation at same time. It is a combination of series controllers connected by a DC link to provide compensation in different transmission lines; for example, interline power flow controller (IPFC). **Combined series-shunt controller** falls in fourth category, which is the combination of a series and shunt FACTS devices, which are connected with a dc link. The DC link enables active power exchange between controllers; for example, unified power flow controller (UPFC) is the combination of SSSC (series FACTS controller) and STATCOM (shunt FACTS controller) which are coupled with a common DC link. It has been observed that designers and researches have proposed various

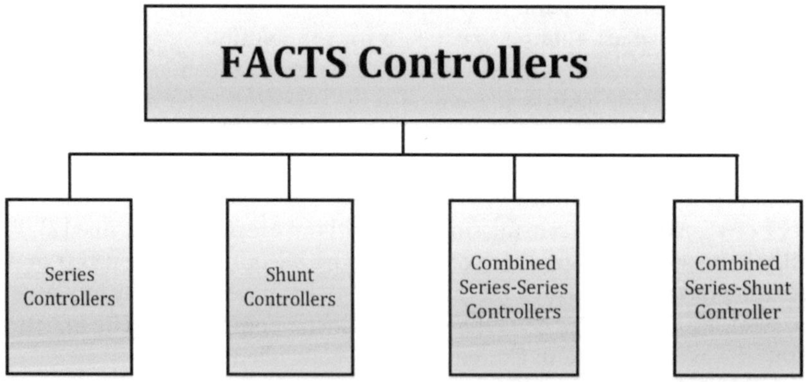

Figure 1.
FACTS controllers.

models for improvement of FACTS device performance with changing power sector scenarios [3–19].

Starting from the conventional power system, a comparative studies of load flow in system interconnection is done between Hessian matrix and Jacobian matrix calculation [3]. The major obstacle in load flow studies have been identified by Burchett et al. [4]. With course of time the power system structure has been changed due to deregulation and independent power producer became the major contributor in power generation [5]. With advancements in power sector policies, conventional methods of active and reactive power dispatch has been changed over the years. So Fuerte had suggested a new approach for optimal power flow problem which was based on Newton's method. This time a new augmented Lagrangian function was associated with the original problem. Implementation of Lagrangian function gave good results as compared to previous ones [6]. Fuerte et al. presented a new and effective method for power flow by incorporating FACTS controllers in power lines. They have also discussed series compensators, phase shifters and tap changing transformers [7]. Pilotto et al. have discussed that how an electricity market has changed after the introduction of FACTS controllers. They have also discussed the rating, location and most efficient types of FACTS controllers for particulate application in power system [8]. Gotham and Heydt presented the model of FACT controllers and also showed that the FACTS devices are solid state converters which have the mechanism for controlling various power system parameters [9].

At the beginning of twenty-first century, Li et al. have developed a steady state model of single FACTS controller for power flow control in the power line. They had proposed genetic algorithm method for efficient power flow control. Result had shown good results in this approach [10]. Canizares discussed the dynamic model of thyristors based FACTS controllers. Models of static VAR compensators, thyristor based series controller, unified power flow controllers had discussed in detail [11]. Dussan P proposed a model which has incorporated the existing load flow method with Newton's method. For designing of static VAR compensators variable shunt susceptance model with Newton's method is used in this case [12]. Yan and Sekar have discussed about the evaluation of rating of FACTS devices. This paper have discussed a whole new framework for designing the alternate power flow network in power system [13]. In years, FACT devices are developed so that it will add contribute more to steady state stability of the system. Keeping this in mind, Biansoongnern proposed the optimum placement of FACTS devices like static VAR compensators (SVC) and thyristor controlled series compensator (TCSC) which will reduce the line losses and increase the power transfer capacity with additional advantage of maintain the voltage profile along the line [14]. Pandya and Joshi has published a paper which emphasis on reduction in generation cost and encourage the use of FACT devices in place of adding another transmission line in the existing system [15]. Hassan and other researchers have proposed a steady state model of series and shunt compensators with firing angle control method. With this proposed model variable power flow has been achieved in the transmission line. With introduction of new generation techniques like renewables, concept of smart grid have introduced in the power system [16]. Smart grid strengthens the transmission and distribution system because of its coordination between various generating source and smart meters installed in consumer's end [17]. The critical technical issues faced by smart grids are also discussed by Colak et al. [18]. With each passing year power system is getting more efficient and stable because of the incorporation of FACTS controllers in smart grid systems [19].

In this chapter the working of SVC and TCSC for OPF is explained. To do so, the chapter is divided into five sections. Section 2 provides the overview of DG and

FACTS in smart grid environment. Section 3 focuses on the optimal power flow solution with SVC and TCSC. Section 4 include the results related to SVC and TCSC using OPF solution. In Section 5, future scope of FACTS devices in smart grid environment is discussed.

2. Overview of DG and FACTS in smart grid environment

Smart grid is the futuristic approach for modernizing the normal grid [2]. Electrical power system is the most complex system which contains all three major sectors, namely, generation, transmission and distribution and were interconnected like one unit, called vertical integrated utility.

In conventional grids, to supply load demands few interconnections were needed among different systems and load shearing between power plants were easy. But, in last few decades, electricity market has grown so fast and there is a need of extra power by different consumers [20]. In order meet increased load demand, many of the generating units are forced to operate at its maximum installed capacity or other solution to get rid from this increased demand of electricity is with the help of distribution generation (DG) [17]. Addition of distribution generation (DG) can make the power grid more reliable in terms of power generation and also can affect the system parameters like voltage or active and reactive power control. Placements of DG and FACTS devices are depicted in **Figure 2**. Due to placement of distributed generator at various points, generation capacity gets

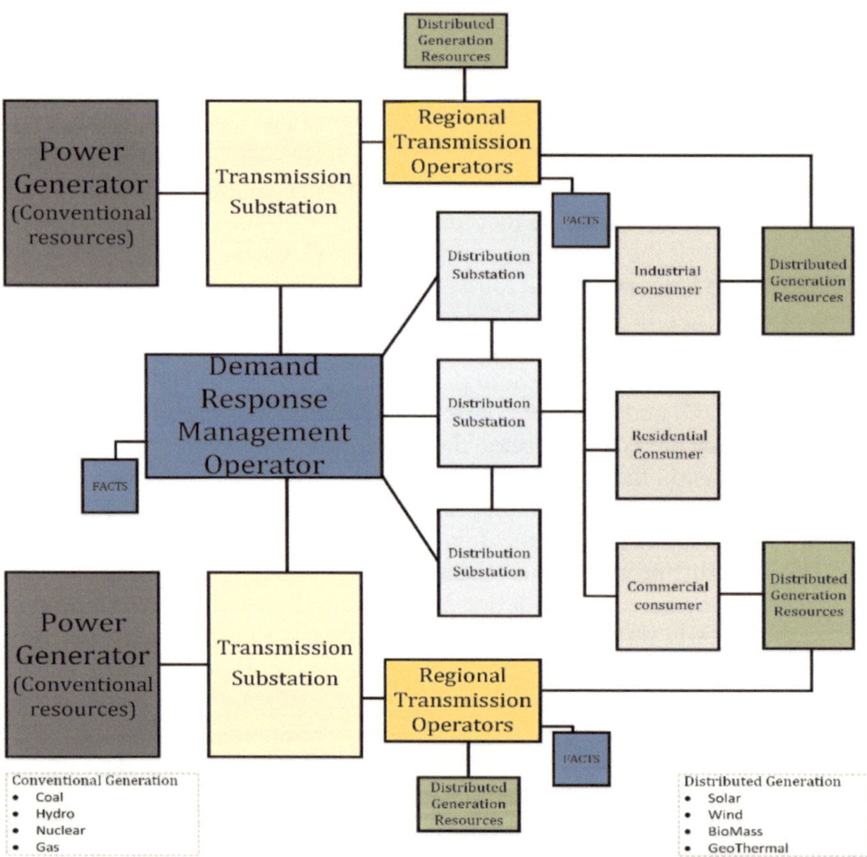

Figure 2.
Smart grid environment using DG and FACTS.

increased and there is a chance of system overload. Also there is an uncertainty of power generation from distributed power generation source and it will lead to under load condition [18]. Situations like overload and underload lead to frequency variation in power system. As discussed earlier, DG increase the generation capacity of the systems will put more stress to the transmission system as there is no other new transmission line for transferring the increase electric energy to various points. It is difficult to erect a new line due to large installation cost. These problems like over voltage, active and reactive power control need to be resolved. Only FACTS devices with power electronics-based control can modify the voltage, phase angle and active power transfer in real time [9, 10]. So FACTS controllers are used for resolving the issues regarding the variation in network parameters imposed by distributed generation. **Figure 2** depicts the smart grid environment with the involvements of DGs and FACTs devices at various levels. Conventional power generation uses coal, nuclear, hydro and gas actuated resources, whereas distributed generation uses solar, wind, biomass, and geothermal resources. Distributed generations may be directly connected to the industrial/commercial users level or it may be connected directly to the transmission level with much higher capacity. Intermittent nature of these renewable energy resources may generate the various problems (as discussed in last paragraph) in the power flow, so in order to get rid of these problems various optimal power flow methods are discussed in the next section.

3. Optimal power flow

Load in power system is distributed in such a manner that each generating unit which is sharing the load will produce electrical power in most economical way. The solution of this economic dispatched is done by optimal power flow method. The idea of OPF concept was first introduced by Carpentier [21]. In OPF method, real and reactive power scheduling is done in such way that the total generation cost gets minimum [22]. It means each power plant is so scheduled that it will generate maximum power with minimum fuel cost. In optimal power flow (OPF) solution, first an objective function is selected (e.g., cost of active power generation). As our power system is highly complex, this objective function is represented by nonlinear equation. Now this objective function is subjected to some system variables and constrains. In Newton's method, the formulation of objective function is much more flexible so that the OPF algorithm can be used for different applications. Newton's method is preferred over other method due to its rapid convergence characteristic near solution [23].

The OPF solution yield various important information about power system and implementation of OPF solution gives more promising results [24].

3.1 Conventional power flow model

The power transfer between sending and receiving end depends on three factors, first one is voltage at each end (shunt compensation), second one is reactance of line (series compensation) and third is phase angle between two ends (phase angle regulation) [25]. These factor can be varied together or separately.

The power transfer P_{km} between two nodes, k and m is

$$P_{km} = \frac{V_k V_m}{X_{km}} \sin(\theta_k - \theta_m) \tag{1}$$

where V_k and V_m are the voltage magnitudes, $\theta_k - \theta_m$ is phase angle difference and X_{km} is the reactance between node k and m.

FACTS controllers are used to modify voltage, line reactance and power angle. It is clear from Eq. (1) that change in above factor will lead to change in power transfer between nodes.

According to Gotham and Heydt, Newton-power flow method was used for estimation of power transfer between generation end and distribution end [9].

In this method, the Jacobian Matrix J is formed and its structure is given below

$$J = \begin{bmatrix} \frac{\partial P}{\partial \theta} & \frac{\partial P}{\partial V} \\ \frac{\partial Q}{\partial \delta} & \frac{\partial Q}{\partial V} \end{bmatrix} = \begin{bmatrix} J_1 J_2 \\ J_3 J_4 \end{bmatrix} \tag{2}$$

The matrix of derivative of state variable describe the power system network in a single equation In AC systems, power system variables can be represented by simultaneous equations:

$$f(X_n AC, R_n F) = 0 \tag{3}$$

$$g(X_n AC, R_n F) = 0 \tag{4}$$

where X_nAC and R_nF are the control variables which stand for reactance and resistance of power system. Dimensions of Jacobian are proportional to numbers and kind of such controller variables.

3.2 Optimal power flow (OPF) concept

The main motive of OPF solution is to meet the load demand while keeping the generation cost minimum. OPF also include economic load dispatch between the generating units by assigning the load to each unit so that the fuel cost as well as losses gets minimum. OPF also maintain system security by maintaining the system in desired operating range at steady state. Maximum and minimum operating range is decided by the operators so that at the time of overload, necessary action can be taken easily. OPF only deals with steady state operating of power system not with transient stability, contingency analysis of power system.

The application of optimal power flow is listed below:

1. OPF is used to calculate optimum generation pattern and to achieve the minimum cost of generation.

2. By using current state of short- and long-term load forecasting, OPF can provide preventive dispatch.

3. At the time of overload, when voltage limits get violated, a corrective dispatch action is provided by optimal power flow solution.

4. OPF is also used to provide optimum generation voltage setting for switched capacitor and for static VAR compensators.

Optimal power flow solution is also used for calculation of bus incremental cost. Bus incremental cost tool is generally used to determine the marginal cost of power at any bus.

3.3 Solution of the optimal power flow

The different methods for solving the OPF are given shown in **Figure 3**.

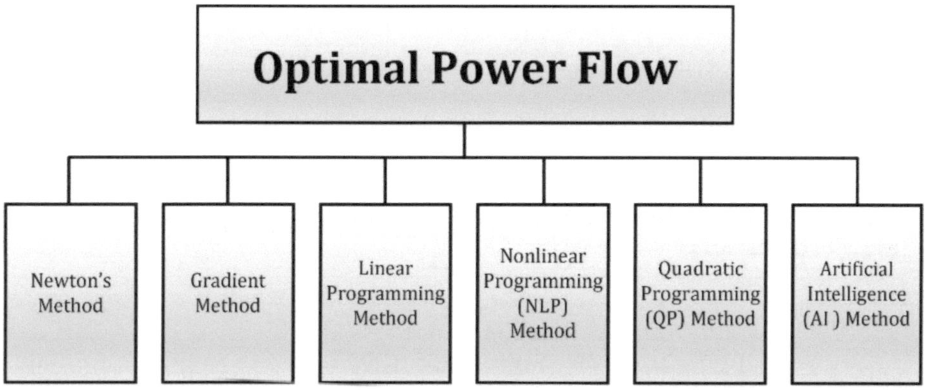

Figure 3.
Optimal power flow methods.

Newton's method (or lambda iteration method) of OPF is preferable among other methods because of its fast convergence. It is the standard method for solving the nonlinear power flow problems. So in coming segments, Newton's method will be discussed for designing of FACTS.

3.4 Newton's method

3.4.1 Variables in Newton's method

First the power flow equations are formulated by Newton's power flow method. After that sparse matrix techniques are applied on power flow equations to attain the solution. In OPF solution, two types of variables are there. First one are control variables and second one depends on control variable called dependent variable. Magnitude of voltage its phase angle and active power at generator buses are considered as controlled variables. Active and reactive power flow (including losses) in all buses except slack bus and phase angle are dependent variable.

3.4.2 Objective function

In OPF solution, the selection of objective function is based on power system economy and power system security. The most commonly used objective function is cost of power generation. For thermal generating units this objective function is generally represented by nonlinear, second order function given below:

$$F = \sum_{i=1}^{Ng} F_i = \sum_{i=1}^{Ng} \left(a_i P_{gi}^2 + b\, P_{gi} + c_i \right) \text{Rs/h} \tag{5}$$

3.4.3 Equality constraints

The power flow equation provides information about the power flow that exists in power network in steady state. For a feasible solution, the power balance must be satisfied, otherwise the OPF solution is referred as infeasible. For a feasible OPF some attempts are being made by relaxing some network constrains which are subjected to

(a) For active power

$$P_i\,(V, \delta) - P_{gi} + P_{di} = 0 \text{ (for } i = 1, 2, \ldots., N_b) \tag{6}$$

(b) For reactive power

$$Q_i\,(V,\delta) - Q_{gi} + Q_{di} = 0 \;(\text{for } i = N_v + 1, N_v + 2, \ldots\ldots.., N_b) \quad (7)$$

3.4.4 Inequality constraints

Variable should satisfy the OPF solution. Inequality constrains define the limits for real power and reactive power generation, voltage magnitude and phase angle. These constraints are:

$$P_{gi}^{min} \le P_{gi} \le P_{gi}^{max} \;(\text{for } i = 1, 2, \ldots., N_b) \quad (8)$$

$$V_i^{min} \le V_i \le V_i^{max} \;(\text{for } i = N_v + 1, N_v + 2, \ldots\ldots.., N_b) \quad (9)$$

$$\delta_i^{min} \le \delta_i \le \delta_i^{max} \;(\text{for } i = 1, 2, \ldots., N_b) \quad (10)$$

$$Q_{gi}^{min} \le Q_{gi} \le Q_{gi}^{max} \;(\text{for } i = N_v + 1, N_v + 2, \ldots\ldots.., N_b) \quad (11)$$

Now the equations of real and reactive power flow is given as

$$P_i(V,\delta) = V_i \sum_{j=1}^{N_b} V_j\left(G_{ij} \cos\left(\delta_i - \delta_j\right) - B_{ij} \cos\left(\delta_i - \delta_j\right)\right) \quad (12)$$

$$Q_i(V,\delta) = V_i \sum_{j=1}^{N_b} V_j\left(G_{ij} \sin\left(\delta_i - \delta_j\right) - B_{ij}\, \text{Sin}\left(\delta_i - \delta_j\right)\right) \quad (13)$$

where V_i, δ_i, P_i, and Q_i are the voltage, phase angle, and active and reactive power for i^{th} bus; P_{gi}, Q_{gi} and P_{di}, Q_{di} are the active and reactive power generation and demand of ith bus and.

N_b, N_g and N_v are is the total no. of buses, generator buses, voltage controlled buses respectively,

In order to form the cost function, power system constrains are incorporated on load flow equation. This new equation with added variable is called incremental cost functions or Lagrange multiplier functions. It can be expressed as:

$$L(P_g, V, \delta V) = F(P_{gi}) + \sum_{i=1}^{N_b} \lambda_{pi}\left(P_i(V, \delta V, \delta)_{gi} + P_{di}\right) + \sum_{i=N_v+1}^{N_b} \lambda_{qi}\left(Q_i(V, \delta V, \delta)_{gi} + Q_{di}\right) \quad (14)$$

After the formulation of cost function, partial derivatives (first and second order) of Lagrangian multiplier (Eq. 14) are calculated with respect to P_{gi}, δ, V_i, λ_{pi}, λ_{qi}.

3.4.5 Penalty function method

With consideration of voltage inequality constrains, an additional function is added to the objective function, F. This additional function is called penalty function, αi. If voltage is outside the limit, the resulting function would be large and the OPF will try to minimize it. Newton's method is second derivative in formation so it is easy to converge. There is only one difficulty which is near the limit, where penalty is small thus it will allow the variable, i.e., if the voltage is above, it will float over its limit. The penalty factor always fulfills the need for inequality constraints. The quadratic penalty factor is described below:

$$\sum_{i=1}^{R} \alpha_i = \sum_{i=1}^{R} \frac{S_i}{2} \left(y_i - \overline{y_i}\right)^2 \quad (15)$$

where $\overline{y_i}$ is for desired value, y_i is the actual value and S_i is the weighting coefficient.

The objective of weighting coefficient is to strengthen the equation. Value of weighting coefficient is calculated by taking the second derivative of penalty factor.

$$\frac{\partial \alpha_i}{\partial y_i} = S_i\left(y_i - \overline{y_i}\right) \tag{16}$$

$$\frac{\partial^2 \alpha_i}{\partial y_i^2} = S_i \tag{17}$$

Value of S_i is varied automatically between two values called hard limit (target limit) and soft limit (lowest value).

3.4.6 Algorithm for OPF based Newton's method

1. Acquire the data a_i, b_i and c_i (for i= total no of generator buses), current load on each bus and total no of buses for existing distribution system.

2. Now compute the Y_{BUS} matrix by the algorithm designed for Y-bus.

3. Compute the initial values of P_{gi} for all generator buses as well as λ by supposing that $P_L = 0$. At that point the problem can be expressed by Eqs. (4) and (5). Hence the solution can be got directly by Eqs. (14) and (2). Initialize all $\lambda_{pi} = 1$, $\lambda_{qi} = 0$, $V_i = 1$p.u. and $\delta_i = 0$.

4. Now compute elements of the Jacobean [J] and Hessian matrix [H], by calculation the first and second order derivatives of Eq. 14

$$[H]\bullet\begin{bmatrix} \Delta P_g \\ \Delta \delta \\ \Delta \lambda_p \\ \Delta V \\ \Delta \lambda_q \end{bmatrix} = -\begin{bmatrix} \frac{\partial L}{\partial P_g} \\ \frac{\partial L}{\partial \delta} \\ \frac{\partial L}{\partial \lambda_p} \\ \frac{\partial L}{\partial V} \\ \frac{\partial L}{\partial \lambda_q} \end{bmatrix} \tag{18}$$

where H is expressed by

$$\begin{bmatrix} \frac{\partial^2 L}{\partial P_g^2} & \frac{\partial^2 L}{\partial P_g \partial \delta} & \frac{\partial^2 L}{\partial P_g \partial \lambda_p} & \frac{\partial^2 L}{\partial P_g \partial V} & \frac{\partial^2 L}{\partial P_g \partial \lambda_q} \\ \frac{\partial^2 L}{\partial \delta \partial P_g} & \frac{\partial^2 L}{\partial \delta^2} & \frac{\partial^2 L}{\partial \delta \partial \lambda_p} & \frac{\partial^2 L}{\partial \delta \partial V} & \frac{\partial^2 L}{\partial \delta \partial \lambda_q} \\ \frac{\partial^2 L}{\partial \lambda_p \partial P_g} & \frac{\partial^2 L}{\partial \lambda_p \partial \delta} & \frac{\partial^2 L}{\partial \lambda_p^2} & \frac{\partial^2 L}{\partial \lambda_p \partial V} & \frac{\partial^2 L}{\partial \lambda_p \partial \lambda_q} \\ \frac{\partial^2 L}{\partial V \partial P_g} & \frac{\partial^2 L}{\partial V \partial \delta} & \frac{\partial^2 L}{\partial V \partial \lambda_p} & \frac{\partial^2 L}{\partial V^2} & \frac{\partial^2 L}{\partial V \partial \lambda_q} \\ \frac{\partial^2 L}{\partial \lambda_q \partial P_g} & \frac{\partial^2 L}{\partial \lambda_q \partial \delta} & \frac{\partial^2 L}{\partial \lambda_q \partial \lambda_p} & \frac{\partial^2 L}{\partial \lambda_q \partial V} & \frac{\partial^2 L}{\partial \lambda_q} \end{bmatrix} \tag{19}$$

Value of ΔP_g, $\Delta\delta$, $\Delta\lambda_p$, ΔV, and $\Delta\lambda_q$ are calculated using Gauss elimination method

5. Checking the convergence and optimal flow conditions. In case the condition violates, GOTO step 6 else GOTO step 8.

$$\left[\sum_{i=1}^{N_g}(\Delta P_{gi})^2+\sum_{i=2}^{N_b}(\Delta\delta_i)^2+\sum_{i=1}^{N_b}(\Delta\lambda_{pi})^2+\sum_{i=N_v+1}^{N_b}(\Delta V_i)^2+\sum_{i=N_v+1}^{N_b}(\Delta\lambda_{qi})^2\right]\leq\in \quad (20)$$

6. After checking the convergence of the problem, values of P_{gi}, δ_i, λ_{pi}, V_i and λ_{qi} has modified as:

$$P_{gi}=P_{gi}+\Delta P_{gi}\ (\text{for } i=1,2,\dots,N_v) \quad (21)$$

$$\delta_i=\delta_i+\Delta\delta_i\ (\text{for } i=2,3,\dots,N_b) \quad (22)$$

$$\lambda_{pi}=\lambda_{pi}+\Delta\lambda_{pi}\ (\text{for } i=1,2,\dots,N_b) \quad (23)$$

$$V_i=V_i+\Delta V_i\ (\text{for } i=N_v+1,N_v+2,\dots,N_b) \quad (24)$$

$$\lambda_{qi}=\lambda_{qi}+\Delta\lambda_{qi}\ (\text{for } i=N_v+1,N_v+2,\dots,N_b) \quad (25)$$

7. For inequality, eliminate penalty factor or equation of power flow. After addition or removal of derivatives, change the equation and now repeat step 4 to update the power flow solution.
8. Compute total generation cost.
9. End.

3.5 An optimal power flow numerical example for 5-bus system

For calculation of bus voltages and generated power, a five-bus network is given in **Figure 4**.

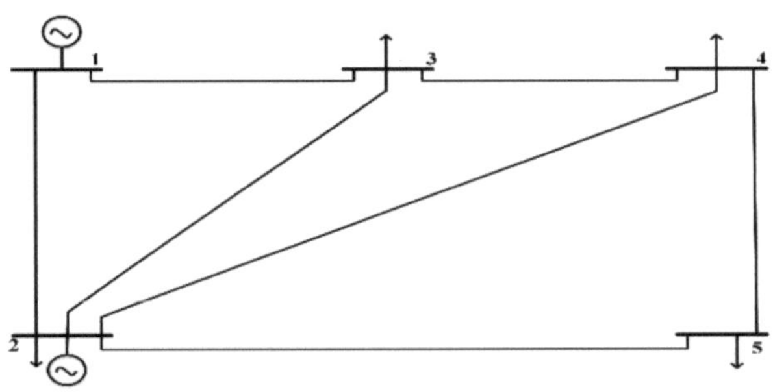

Figure 4.
A typical 5-bus system.

Bus no.	V (p.u.)	δ (rad)	λ_p
1	1.06	0	3.4052
2	1.00	−0.0014	3.4084
3	0.9875	−0.0555	3.5437
4	0.9843	−0.0593	3.5520
5	0.9717	−0.0684	3.5760

Table 1.
Node parameters for the 5-bus system.

Line no.	From bus	To bus	Ps (p.u.)	Qs (p.u.)	Pr (p.u.)	Qr (p.u.)	PLoss (p.u.)	QLoss (p.u.)
1	1	2	0.4077	j0.9466	−0.2623	−j0.8926	0.1454	j0.0540
2	1	3	0.5855	j0.2217	−0.0712	−j0.1899	0.5143	j0.0318
3	2	3	0.4702	−j0.0193	−0.1096	−j0.0345	0.3606	j0.0152
4	2	4	0.4937	−j0.0082	−0.1334	−j0.0259	0.3603	j0.0177
5	2	5	0.6841	j0.0660	−0.4379	−j0.0273	0.2462	j0.0387
6	3	4	0.1717	j0.0581	−0.1132	−j0.0573	0.0586	j0.0007
7	4	5	0.2807	j0.0358	−0.0349	−j0.1787	0.4594	j0.0009

Table 2.
Line flows for the 5-bus system.

Tables 1 and **2** shows the resulting power flows, the node voltages and values of Lagrange multipliers at operating points.

3.6 OPF analysis using FACTS

From literature survey done on FACTS controller, it has been observed that only few research papers are there for the modeling of FACTS-devices. Yan and Sekar proposed a mathematical model for TCSC, IPC and UPFC [13]. Also, in Handschin and Lehmkoester research paper, an improvisation has done in the modeling field and mathematical models of UPFC, SSSC have been proposed using OPF solution [24]. The next section covers series (TCSC) and shunt (SVC) FACT devices.

3.7 Static VAR compensator (SVC)

SVC is a shunt FACTS devices used to maintain voltage profile along the transmission line. The word "static" stands for the device with no moving parts. It is a shunt connected device consists of thyristor switches with an assembly of inductors and capacitors connected on parallel. This shunt FACTS device absorb or inject the reactive as well as active power in the system. **Figure 5** shows the basic SVC diagram.

3.7.1 OPF incorporating SVC

In this section discuss the designing methods of SVC using Newton's method of OPF. In this method it has been assumed that in order fulfill target voltage requirement, SVC should act as a variable shunt susceptance. Two type of SVC designing methods are described in this section.

a. Shunt susceptance method

b. Firing angle control method

3.7.2 Lagrangian function

Lagrangian function for SVC is formulated by transforming the constrained power flow equation into unconstrained one. Lagrangian function is denoted by L (x,λ) which is the summation of objective function $f(P_g)$ and product of Lagrange multiplier vector λ and power flow equation $[P_g,V,\theta, B(\alpha)]$.

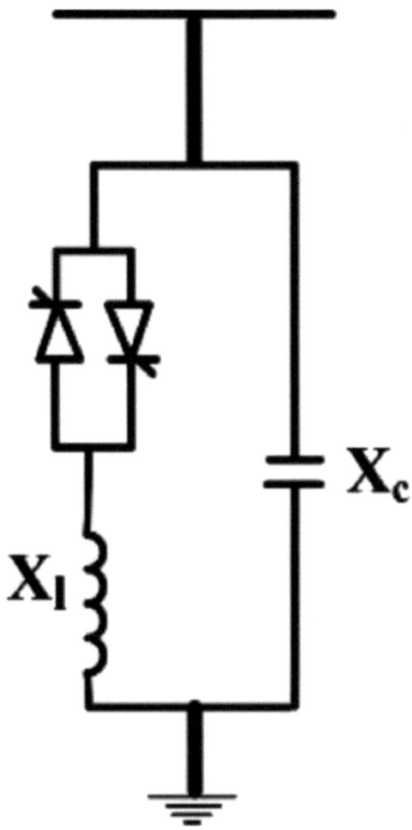

Figure 5.
Static VAR compensator (SVC).

And shunt susceptance model of SVC is expressed as.

$$L(x,\lambda) = f(P_g) + \lambda^{th}[P_g, V, \theta, B(\alpha)] \tag{26}$$

where P_g is generated active power x is variable vector; $B(\alpha)$ is the shunt susceptance of SVC. In above case only equality constraints are considered.

So keeping the equality constraints in consideration, the Lagrangian function formulated for SVC is given as.

$$L_{SVC}(x,\lambda) = \lambda_{qk} Q_k \tag{27}$$

$$Q_k = -V_k^2 B_{svc} \tag{28}$$

where designing of SVC can be done with two different methods. These methods will be discussed in next section.

3.7.3 SVC total susceptance model (B = BSVC)

In total susceptance model of SVC, value of fundamental component of susceptance B_{svc} and equivalent reactance of SVC is calculated. The TCR is the combination of capacitor and inductance with a bi-directional Thyristors valve which is attached in parallel with a fixed capacitor.

At fundamental frequency, the equivalent reactance, X_{Leq}

$$X_{Leq} = \frac{\pi X_L}{\sin 2\alpha + 2(\pi - \alpha)} \tag{29}$$

where α is the thyristor firing angle. X_L is inductive reactance, and X_c is capacitive reactance.

Value of SVC susceptance is

$$B_{SVC} = \frac{X_L - \frac{X_C}{\pi}(2(\pi - \alpha)) + \sin(2\alpha)}{X_C X_L} \tag{30}$$

And value of reactive power is:

$$Q_i^{SVC} = -V_i^2 B_{SVC} \tag{31}$$

After the computation of B_{SVC} and Q_i^{SVC}, elements of the Jacobean (Eq. 18) and Hessian matrix (Eq. 19) will be calculated by taking first and second order derivatives of Eq. 14.

3.7.4 SVC firing angle control model

In TCR-FC configuration of SVC, TCR branch has a reactor in series with thyristor pair. Its inductive reactance (X_L) can be controlled by changing the firing angle α. Because of this, total reactance of the SVC (X_{SVC}) get changed as reactor is in parallel with fixed capacitor. So the susceptance B_{SVC} is given as

$$I_{SVC} = -jB_{SVC}V_k \tag{32}$$

And the reactance of TCR is given by the formula

$$X_{TCR} = \frac{\pi X_L}{\sigma - \sin\sigma} \tag{33}$$

Now after putting the value $\sigma = 2(\pi\text{-}\alpha)$

$$X_{TCR} = \frac{\pi X_L}{2(\pi - \alpha) + \sin(2\alpha)} \tag{34}$$

where σ and α are conduction and firing angles, respectively.

In SVC, TCR is in parallel with capacitor so, total reactance of SVC will be $(X_L \| X_C)$

$$X_{SVC} = \frac{\pi X_C X_L}{X_C[2(\pi - a) + \sin 2a] - \pi X_L} \tag{35}$$

where $X_c = 1/wC$,

After the computation of B_{SVC}, elements of the Jacobean (Eq. 19) and Hessian matrix (Eq. 20) will be calculated by taking first and second order derivatives of Eq. 13.

3.7.5 Lagrange multiplier

In case of SVC, Lagrange multiplier is initialized at $\lambda_{pk} = 1$ and $\lambda_{qk} = 0$.

3.7.6 OPF test cases for SVC

The above defined method for SVC has implemented in an OPF to test algorithm. Testing is done on 5-bus system discussed in Section 3.5. The objective function is active power generation cost. **Figure 6** shows a 5-bus system incorporating SVC.

Tables 3–6 shows the Node voltage, optimal generation cost of a standard 5-bus system for susceptance model and firing angle control model respectively.

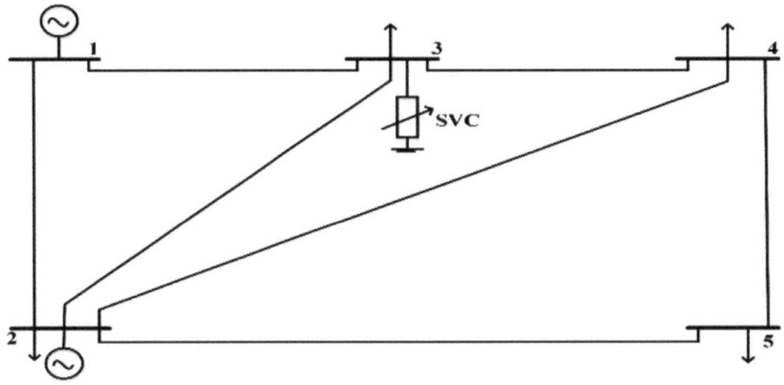

Figure 6.
A typical 5-bus system incorporating SVC.

Bus no.	V (p.u.)	δ (rad)	λp
1	1.06	0	3.4052
2	1.00	−0.0013	3.4084
3	1	−0.0629	3.5519
4	1.0053	−0.0649	3.5600
5	0.9788	−0.0699	3.5780

Table 3.
Node parameters for the 5-bus system for susceptance model of SVC.

Line no.	From bus	To bus	Ps (p.u.)	Qs (p.u.)	Pr (p.u.)	Qr (p.u.)	PLoss (p.u.)	QLoss (p.u.)
1	1	2	0.4061	j0.9471	−0.2607	−j0.8931	0.1454	j0.0540
2	1	3	0.5870	j0.1076	−0.0627	−j0.0836	0.5243	j0.0240
3	2	3	0.4724	−j0.1629	−0.1007	j0.1831	0.3717	j0.0202
4.	2	4	0.4940	−j0.1228	−0.1253	j0.1433	0.3687	j0.0205
5	2	5	0.6820	j0.0285	−0.4344	j0.0294	0.2476	j0.0379
6	3	4	0.1771	j0.2351	−0.1153	−j0.2329	0.0619	j0.0022
7	4	5	0.2943	j0.0938	0.1791	−j0.0911	0.4734	j0.0027

Table 4.
Line flows for the 5-bus system for susceptance model of SVC.

Bus no.	V (p.u.)	δ (rad)	λp
1	1.06	0	3.4053
2	1.00	−0.0014	3.4084
3	1.00	−0.0582	3.5403
4	0.9917	−0.0613	3.5485
5	0.9742	−0.0690	3.5739

Table 5.
Nodal parameters for the 5-bus system for firing angle model of SVC.

Line no.	From bus	To bus	Ps (p.u.)	Qs (p.u.)	Pr (p.u.)	Qr (p.u.)	PLoss (p.u.)	QLoss (p.u.)
1	1	2	0.4077	j0.9466	−0.2623	−j0.8926	0.1454	j0.0540
2	1	3	0.5826	j0.1682	−0.0640	−j0.1412	0.5187	j0.0270
3	2	3	0.4665	−j0.0866	−0.1012	−j0.1027	0.3654	j0.0161
4	2	4	0.4936	−j0.0486	−0.1306	−j0.0667	0.3631	j0.0181
5	2	5	0.6836	−0.0457	−0.4369	−j0.0073	0.2467	j0.0384
6	3	4	0.2053	j0.2184	−0.1450	−j0.2160	0.0603	j0.0024
7	4	5	0.2857	j0.0559	0.1786	−j0.0545	0.4643	j0.0014

Table 6.
Line flows for the 5-bus system for firing angle model of SVC.

3.8 Thyristor-controlled series capacitor (TCSC)

Thyristor control series capacitor is used to provide the variable impedance to the network. TCSC assembly consists of a combination of capacitor in parallel with Thyristor controlled reactor. The overall reactance of TCSC is the parallel combination of capacitive reactance and variable inductive reactance. By changing the reactance of the line, this controller will change the power transfer capacity of the line. Diagram of TCSC is shown in **Figure 7**.

3.9 OPF incorporating TCSC

In this section the OPF TCSC designing is done. TCSC model is adjusted according to the Newton's method for OPF calculation. For designing, it has been assumed that the series reactance of TCSC is the non-linear function of firing angle. Lagrangian function for TCSC is formulated by transforming the constrained power flow equation into unconstrained one. Lagrangian function is denoted by L(x,λ) which is the summation of objective function f(Pg) and product of Lagrange

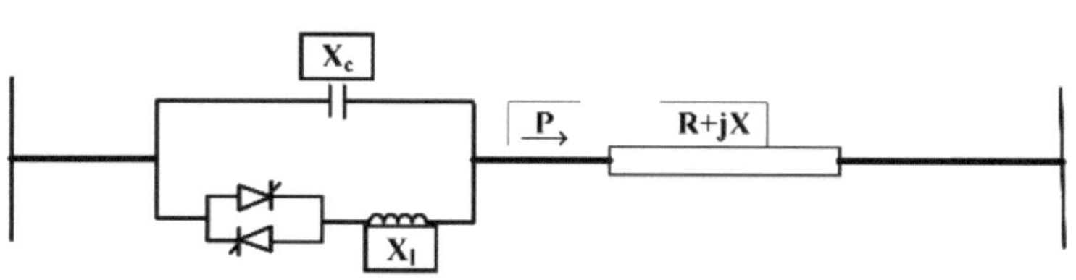

Figure 7.
Thyristor controlled series capacitor.

multiplier vector λ and power flow equation [Pg,V,θ, B(α)]. And shunt susceptance model of SVC is expressed as.

$$L(x, \lambda) = f(P_g) + \lambda^{th}[P_g, V, \theta, B(\alpha)] \tag{36}$$

where P_g is generated active power; B(α) is the shunt susceptance of SVC. In above case only equality constraints are considered.

The power flow equations for bus k and m using Lagrangian function can be written as

$$\begin{aligned} L_{tcsc.}(x, \lambda) &= \lambda_{pk}(P_k + P_{dk} - P_{gk}) + \lambda_{qk}\left(Q_k + Q_{dk} - Q_{gk}\right) \\ &+ \lambda_{pm}(P_m + P_{dm} - P_{gm}) + \lambda_{gm}\left(Q_m + Q_{dm} - Q_{gm}\right) \end{aligned} \tag{37}$$

where values λ of are for Lagrange multipliers for k and m bus; P_d, Q_d are the load demand for bus k and m. and Pg, Qg are the scheduled power generation. For branch k-l the Lagrangian function, L

$$L = L_{tcsc.}(x, \lambda) + L_{flow}(x, \lambda) \tag{38}$$

where $L_{flow} = \lambda_{ml}(P_{ml}\text{-}P_{specified})$ and λ_{ml} is Lagrange multiplier for active power flow in branch *m-l*.

Reactance of TCSC is given by

$$X_{TCSC}(\alpha) = \frac{X_C X_{L\,(\alpha)}}{X_{L\,(\alpha)} - X_L} \tag{39}$$

$$X_L = X_L(\pi/\pi - 2\alpha - \sin 2\alpha) \tag{40}$$

$$X_{TCSC}(\alpha) = \frac{X_L X_C}{\frac{X_C}{\pi}\,[2(\pi - \alpha) + \sin 2\alpha] - X_L} \tag{41}$$

3.10 TCSC test case

Designing of TCSC can be done with two different methods. First one is variable reactance model and second is firing angle control model of TCSC. For both models, XTCSC, and BTCSC are calculated using Lagrange multiplier. After the computation of B_{TCSC}, and XTCSC, elements of the Jacobean (Eq. 15) and Hessian matrix (Eq. 18) will be calculated by taking derivatives of Eq. (13). Objective function for TCSC is cost of active power generation. After designing, variable impedance TCSC model and firing angle control TCSC model will be tested on 5-bus system. A 5-bus

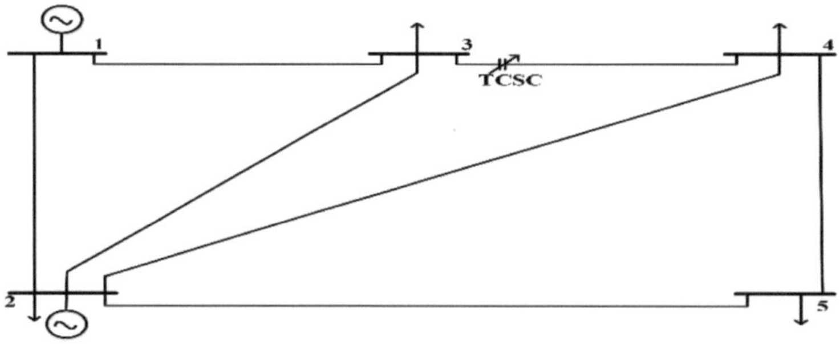

Figure 8.
A typical 5-bus system incorporating TCSC.

Bus no.	V (p.u.)	δ (rad)	λp
1	1.06	0	3.4641
2	1.00	−0.0065	3.5023
3	1.048	−0.0579	3.5387
4	0.977	−0.0522	3.4581
5	0.972	−0.0611	3.5519

Table 7.
Node parameters for the 5-bus system for variable impedance model.

Line no.	From bus	To bus	Ps (p.u.)	Qs (p.u.)	Pr (p.u.)	Qr (p.u.)	PLoss (p.u.)	QLoss (p.u.)
1	1	2	0.4889	j0.9199	−0.3432	−j0.8652	0.1456	j0.0547
2	1	3	0.5290	−j0.0257	0.0091	j0.0402	0.5381	j0.0145
3	2	3	0.3715	−j0.3228	0.0146	j0.3482	0.3861	j0.0254
4	2	4	0.4432	j0.0457	−0.0871	−j0.0329	0.3561	j0.0128
5	2	5	0.5915	j0.0883	−0.3489	−j0.0606	0.2426	j0.0276
6	3	4	0.6021	j2.2911	−0.4898	−j2.1389	0.1123	j0.1522
7	4	5	0.2669	j0.0079	0.1890	−j0.0075	0.4560	j0.0004

Table 8.
Line flows for the 5-bus system for variable impedance model.

Bus no.	V (p.u.)	δ (rad)	λp
1	1.06	0	3.4541
2	1.00	−0.024	3.4632
3	1.036	−0.0613	3.4562
4	0.962	−0.0642	3.4431
5	0.971	−0.0653	3.5562

Table 9.
Nodal parameters for the 5-bus system for firing angle model of TCSC.

Line no.	From bus	To bus	Ps (p.u.)	Qs (p.u.)	Pr (p.u.)	Qr (p.u.)	PLoss (p.u.)	QLoss (p.u.)
1	1	2	0.7685	j0.8314	−0.6200	−j0.7682)	0.1485	j0.0632
2	1	3	0.5563	j0.0190	−0.0232	−j0.0014)	0.5331	j0.0176
3	2	3	0.3144	−j0.2498	0.0634	j0.2545	0.3778	j0.0137
4	2	4	0.4379	j0.1294	−0.0864	−j0.1145)	0.3516	j0.0150
5	2	5	0.4953	j0.1235	−0.2559	−j0.1048)	0.2394	j0.0187
6	3	4	0.8856	j2.2711	−0.7708	−j2.1066)	0.1148	j0.1645
7	4	5	0.2151	−j0.0330	0.2334	j0.0341	0.4485	j0.0003

Table 10.
Line flows for the 5-bus system for firing angle model of TCSC.

system incorporating TCSC is shown in **Figure 8**. **Tables 7–10** shows the nodal parameters and optimal generation cost for modified with system losses for variable impedance model and firing angle control modal respectively.

4. Result and discussion

For standard 5 bus system, the resulting power flows, node voltages are given in **Tables 1** and **2**, the generation cost and active power losses are presented in tabular form in **Table 11**.

Quantity	Value
cost of Active power generation	127.59 Rs/h
Active power generation	1.7031 p.u.
Reactive power generation	0.400 p.u.

Table 11.
OPF solution for the 5-bus system by Newton's method.

Above table shows that all nodal voltages edge toward high voltage side. The purpose of OPF solution is served as the multiplier method handled the limit violation efficiently which happed during the iteration process. Also the power production by two generators after optimal power flow solution is different from conventional solution. In conventional solution, an undesirable situation arise as there is a mismatched generation of power for both generator used in standard 5 bus system. But in OPF solution, the produced and absorbed reactive power is function of optimization algorithm which enable each generator to hold the even share of active power requirement.

4.1 The static VAR compensator (SVC)

The OPF solution for new model of SVC are tested on standard 5-bus system which are used previously. **Tables 12** and **13** provide the detail of optimal generation cost for Upgraded SVC with susceptance model and firing angle control model.

Quantity	Value
cost of Active power generation	125.78 Rs/h
Active power generation	1.7037 p.u.
Reactive power generation	0.5460 p.u.

Table 12.
OPF solution for the 5-bus system by using susceptance model of SVC.

Quantity	Value
Cost of Active power generation	125.75 Rs/h
Active power generation	1.7024 p.u.
Reactive power generation	0.5532 p.u.

Table 13.
OPF solution for the 5-bus system by using firing angle model of SVC.

The above table shows that the generation cost is reduced after the implementation of SVCs and the voltage profile is also improved. So it is clear that incorporation of SVC in the system will lead to content voltage profile along the line.

On comparing above two cases, it can be concluded that the generation cost is almost equal. In early iteration, there are oscillation in cost and losses due to variation in penalty weighing factor.

4.2 Thyristor-controlled series compensator (TCSC)

Like SVCs, the OPF solution for new model of TCSC is tested on, Standard 5-bus system which were used previously. The detail of optimal generation cost for upgraded SVC is provided by **Tables 14** and **15**.

Quantity	Value
Cost of Active power generation	125.72 Rs/h
Active power generation	1.6952 p.u.
Reactive power generation	0.4324 p.u.

Table 14.
OPF solution for the 5-bus system by using susceptance model of TCSC.

Quantity	Value
Cost of Active power generation cost	125.718 Rs/h
Active power generation	1.7002 p.u.
Reactive power generation	0.4623 p.u.

Table 15.
OPF solution for the 5-bus system by using firing angle model of TCSC.

On comparing both models of OPF it can be observed that, the generation cost is a bit different but both models work in favor of increase in power transfer capacity of transmission line.

5. Conclusion

This chapter includes the optimum power flow solution for FACTS devices in smart grid environment. FACTS devices functions are evolving from simply sustaining the stability of the transmission system to increasing of power transfer capability hence improvement of overall performance of transmission line. It can be concluded that over the year, various optimization techniques, numerical methods are used for solving the optimum power flow problems. In today's scenario, currently available OPF algorithm satisfy all the full nonlinear load flow model and its boundary variables. Newton's method is one of the newest OPF algorithm and gives highest convergence characteristic. Since 1980, there are so many improvements in numeric techniques and introduction of computer based numerical techniques have given it a tremendous exposure. But even after such a remarkable advancement, the OPF solution is a difficult mathematical problem to solve. In real time OPF is more complex nonlinear problem which are subjected to real time constrains, and sometime prone to some ill real time conditions and difficult to converge. So a new OPF algorithm can be recommended for future which will have the ability to overcome the drawback that is encountered in real time application.

5.1 Future scope

From several years thyristor-based phase controlled switches has been used for FACTS device application, and considered as conventional. Now a days, more promising switch-mode GTO-based switches are introduces in place of conventional thyristor switch. A hybrid approach involving both thyristor and GTO based switches are suggested for future FACTS controller.

The ongoing restructuring of power system leads to power system stability problem because of the change in power transfer patters between generation, transmission and distribution companies. FACTS controllers are used for accomplishing stability objectives. OPF algorithm for FACTS controller ensure the placement of FACTS devices in such a manner that will ensure system stability, content voltage profile and improved overall reliability of the power system.

Author details

Ankur Singh Rana[1*], Mohit Bajaj[2] and Shrija Gairola[3]

1 National Institute of Technology Tiruchirappalli, Tamil Nadu, India

2 National Institute of Technology Delhi, New Delhi, India

3 THDC Institute of Hydropower Engineering and Technology, Uttarakhand, India

*Address all correspondence to: ankurranag@gmail.com

References

[1] Pradeep Y, Karapade SA, Kumar R. Intelligent grid initiative in India. In: Proceedings of 14th IEEE International Conference on Intelligent System Application to Power System (ISAP). IEEE; 5-8 Nov. 2007. pp. 1-6. DOI: 10.1109/ISAP.2007.4441670

[2] Mahmoud MS, Xia Y. Chapter 7— Smart Grid Infrastructures, Networked Control Systems. Butterworth-Heinemann; 2019. pp. 315-349

[3] Sasson AM, Viloria F, Aboytes F. Optimal load flow solution using the Hessian matrix. IEEE Transactions on Power Apparatus and Systems. 1973; **95**(1):31-41

[4] Burchett RC, Happ HH, Vierath DR, Wirgau KA. Developments in optimal power flow. IEEE Transactions on Power Apparatus and Systems. 1982; **101**(2):406-414

[5] Campbell MK. Power system degregualtion. IEEE Potentials. 2002; **20**(5):8-9

[6] Ambriz PH, Acha E, Fuerte ECR. Advanced SVC models for Newton-Raphson load flow and Newton optimal flow studies. IEEE Transactions on Power Systems. 2000;**15**(1):129-136

[7] Fuerte ECR, Acha E. Newton-Raphson algorithm for the reliable solution of large power networks with embedded FACTS devices. IEEE Proceedings on Generation, Transmission and Distribution. 1996;**143**(5):447-454

[8] Pilotto LAS, Ping WW, Carvalho AR, Wey A, Long WF, Alvarado FL, et al. Determination of need FACTS controllers. IEEE Transactions on Power Delivery. 1997;**12**(1):364-371. DOI: 10.1109/61.568260

[9] Gotham DJ, Heydt GT. Power flow control and power flow studies for systems with FACTS devices. IEEE Transactions on Power Systems. 1998;**13**(1):60-65. DOI: 10.1109/59.651614

[10] Li N, Yan X, Heng C. Facts-based power flow control in interconnected power systems. IEEE Transactions on Power Systems. 2000;**15**(1):257-262

[11] Canizares CA. Power flow and transient stability models of FACTS controllers for voltage and angle stability studies. In: IEEE; 2000 IEEE Power Engineering Society Winter Meeting. Conference Proceedings (Cat. No.00CH37077). IEEE PES WM Panel on Modeling, Simulation and Applications of FACTS Controllers in Angle and Voltage Stability Studies; Singapore; 2000

[12] Dussan P. Modelling of FACTS in power system studies. IEEE; 2000 IEEE Power Engineering Society Winter Meeting. Conference Proceedings (Cat.No.00CH37077). Vol. 2. Singapore: IEEE; 2000. pp. 1435-1439. DOI: 10.1109/PESW.2000.850190

[13] Yan P, Sekar A. Steady-state analysis of power system having multiple FACTS devices using line-flow-based equations. IEEE Proceedings on Generation, Transmission and Distribution. 2005; **152**(1):31-39. DOI: 10.1049/ip-gtd: 20041133

[14] Biansoongnern S, Chusanapiputt S, Phoomvuthisarn S. Optimal SVC and TCSC placement for of transmission losses. In: Proceedings of the International Conference on Power Syatem Technology; 2006. Correct paper title: Optimal SVC and TCSC Placement for Minimization of Transmission Losses. IEEE; 2006 International Conference on Power System Technology, 22-26 Oct. 2006. DOI: 10.1109/ICPST.2006.321926

[15] Pandya KS, Joshi SK. A survey of optimal power flow methods. Journal of Theoretical and Applied Information Technology. 2008;**4**:450-458

[16] Murthy VKS, Kharparde SA, Gupta RP, Pradeep Y. Smart grid initiative and power market in India. In: Proceeding IEEE Power and Energy Society General Meeting. IEEE; July 2010. pp. 1-7. DOI: 10.1109/PES.2010.5589947

[17] Bayindir R, Cola I, Fulli G, Demirtas K. Smart grid technologies and applications. Renewable and Sustainable Energy Reviews. 2016;**66**:499-516

[18] Colak I, Sagiroglu S, Fulli G, Yesilbudak M. A survey on the critical issues in smart grid technologies. Renewable and Sustainable Energy Reviews. 2016;**54**:396-405

[19] Gandoman FH, Ahmadi A, Sharaf AM, Siano P, Pou J, Hredzak B, et al. Review of FACTS technologies and applications for power quality in smart grids with renewable energy systems. Renewable and Sustainable Energy Reviews Part 1. 2018;**82**:502-514

[20] Ni Y, Zhong J, Liu H. Deregulation of power system in Asia: Special consideration in developing countries. In: Proceedings of the IEEE Power Engineering Society General Meeting; June 2005. IEEE; IEEE Power Engineering Society General Meeting. Vol. 3. 16-16 June 2005. pp. 2876-2881. DOI: 10.1109/PES.2005.1489411

[21] Carpentier JL. Optimal power flow: Uses, methods and developments. Proceedings of IFAC Conference. 1985; **18**(7):11-21. DOI: 10.1016/S1474-6670 (17)60410-5

[22] Hassan MO, Cheng SJ, Zakaria A. Steady-state modeling of SVC and TCSC for power flow analysis. In: Proceedings of the International Multi Conference of Engineers and Computer Scientists. Vol. 2. IMECS; 2009

[23] Gotham DJ, Heydt GT. Power flow control and power flow studies for systems with FACTS device. IEEE Transactions on Power Systems. 1998;**13**(1):60-65

[24] Handschin E, Lehmkoester C. Optimal power flow for deregulation systems with FACTS-devices. In: Power Systems Computation Conference (PSCC); 13th Power Systems Computation Conference Proceedings: Trondheim, Norway, June 28-July 2nd 1999. 13th PSCC; Trondheim, Norway; 1999. pp. 1270-1276

[25] Hingorani NG, Gyugyi L. Understanding FACTS: Concept and Technology of Flexible AC Transmission System. Wiley-IEEE Press; 2000

Chapter 3

Reducing Power Losses in Smart Grids with Cooperative Game Theory

Javier B. Cabrera, Manuel F. Veiga, Diego X. Morales and Ricardo Medina

Abstract

In a theoretical framework of game theory, one can distinguish between the noncooperative and the cooperative game theory. While the theory of noncooperative games is about modeling competitive behavior, cooperative game theory is dedicated to the study of cooperation among a number of players. The cooperative game theory includes mostly two branches: the Nash negotiation and the coalitional game theory. In this chapter, we restrict our attention to the latter. In recent years, the concept of efficient management of electric power has become more complex as a result of the high integration of distributed energy resources in the scenarios to be considered, mainly distributed generation, energy storage distributed, and demand management. This situation has been accentuated with the appearance of new consumption elements, such as electric vehicles, which could cause a high impact on distribution gridworks if they are not managed properly. This chapter presents an innovative approach toward an efficient energy model through the application of the theory of cooperative games with transferable utility in which the management, capacity, and control of distributed energy resources are integrated to provide optimal energy solutions that allow achieving significant savings in associated costs. This chapter presents a general description of the potential of the application of the theory to address Smart Grid, providing a systematic treatment.

Keywords: game theory, coalition, cooperative, Smart Grid, power loss

1. Introduction

Electricity consumption has grown in terms of the advances in technology, but we must bear in mind that this demand for electricity is variable at different times of the day. It is therefore possible to divide a day into two parts, namely, the maximum and minimum demand periods [1]. For 1 day, the maximum demand consists of the most active time of electricity consumption, and the maximum demand differs depending on the season. If power plants are able to consistently maintain high power generation, they can meet the maximum demand. However, the high production of electricity, especially obtained from nonrenewable energy resources (e.g., thermoelectric power plants), usually wastes a lot of energy. Therefore, we require a new type of intelligent electrical grid, which can help power plants to be more efficient, reliable, and solid, to avoid the generation of unnecessary energy and/or loss of energy in the distribution.

Microgrids (MGs) comprising distributed power generators have been introduced recently to construct smart grid to reduce power loss. MGs are able to supply electricity to the end users (i.e., homes, companies, schools, and so forth) which are linked to the corresponding MGs [2]. The MGs can exchange power with others. In addition, they are also capable of transferring power with the macro station (MS), which is the primary substation of the smart grid. In the presence of MGs, it is desirable to allow the microgrids to service some small geographical areas or group of customers based on their demand, so as to relieve the demand on the main grid [2]. We consider a power network consisting of interconnected microgrids and a macrogrid. The MGs harvest renewable energy (e.g., wind, solar, etc.), whereas the macrogrid produces energy from conventional sources. The MGs are equipped with storage devices (e.g., batteries) in which they can store energy for future usage locally. Although these resources are easily procurable and depicted as "green" energy resources, they present a significant shortcoming since they cannot guarantee stable production of electricity at all times [3]. For example [4], solar energy generation through deployed solar panels in the MGs can be seriously hampered on rainy days. When a MG needs additional power, it can buy electricity from the wholesaler (i.e., the MS) and/or from neighboring MGs.

Kantarci et al. proposed the "cost-aware smart microgrid network design," which enables economic power transactions within the smart grid [5, 6]. The problem of power loss minimization was discussed in the work conducted by Meliopoulos et al. [7, 8] whereby a real-time and coordinated control scheme was proposed with the participation of distributed generation resources that can be coordinated with the existing infrastructure [9–11].

Kirthiga et al. proposed a detailed methodology to develop an autonomous microgrid for addressing power loss in [12]. Furthermore, some researchers have addressed power loss in the works in [13–15].

At present, game theory is an important tool for microgrid research as described in the work in [16–18]. Saad et al. presented an algorithm based on the cooperative game theory to study novel cooperative strategies between the microgrids of a distribution network [19].

The challenge of the electric companies is to determine the mechanisms that allow efficiently and quickly the equal distribution of the electric power surrendered by the electricity distribution grid as well as the distributed generation and that the clients or consumers of that energy have a common benefit.

According to the energy current pattern, the chain of the use of the energy was based on the generation stages, transport, distribution-commercialization, and consumption. This model in some countries differs basically in the form of the electric market, that is to say, in countries like Ecuador, Venezuela, and Mexico, the market structure is monopolist which has a single company constituted by subcompanies denominated as generation company, transmission company, and distribution companies. The price for the energy is fixed by the institutions of the State that regulate the electric sector. In other countries, mainly European countries, the market pattern is based on the free offer on the part of the generation companies, consumers can choose the company freely to which they want to buy the product, and the transmissions and distribution companies allow to carry out these transactions acting as intermediaries in the energy sale. From a general perspective, it is foreseen that the new smart electric grid is a cyber-physical system of a large scale that can improve the efficiency, dependability, and robustness of the electric grids, by means of the integration of advanced techniques, as control, communications, and signal processing. Intrinsically, the smart electric grid is an energy grid made up of intelligent nodes that can operate, communicate, and interact, in an autonomous way, to provide efficient electrical power to its consumers. The heterogeneous nature of the smart electric grid motivates the adoption of advanced techniques to overcome the diverse technical challenges in different levels as the design, control, and implementation.

In this sense, it is expected that the theory of games constitutes an essential analytic tool in the design of the future smart power grid, as well as in the cyber-physical systems to a large scale. The theory of games is a formal framework as much analytic as conceptual with a group of mathematical tools that allow the study of complex interactions among rational, independent players.

2. Electric system model for a cooperative game

Considering a single macro station denominated by a transmission substation, this macro station has a group of N Smart Grid, which a certain period of time can behave as microgrids that have an energy surplus (sellers) or energy requirements (buyers). Thus, a coalition formed in the grid can have any of these two types of Smart Grid.

One of the initial hypotheses to consider the exchange pattern based on a cooperative game is that all the Smart Grid possesses the information of the grid that allows choosing one of them. Being part of a specific coalition is always know, and the link between all and each one of Smart Grid belonging to the certain Macro station is always feasible, having as a result that all the members of the electric grid can interact with each other.

A specific electric grid may be made up of a group of Smart Grid, where for the i-th Smart Grid in a particular frame of time it can be said that this microgrid has a generated total power called P_i and at the same time a power demand by a group of consumers that is shown in D_i. Therefore, the surplus power to the Smart Grid $i \in N$ is given by [20]:

$$Q_i = P_i - D_i \tag{1}$$

Depending on the power generation values and electrical demand in Smart Grid, the surplus energy can define three cases to analyze:

- Case 1: $Q_i > 0$.In this case, the Smart Grid has a surplus power which makes it able to sell this electric power (seller) and shaping coalitions with the Smart Grid or substation.
- Case 2: $Q_i = 0$. In this case, the Smart Grid supplies its consumption.
- Case 3: $Q_i < 0$.Here the Smart Grid can buy electric power (buyer) from another Smart Grid or substation.

It should be kept in mind that both the power generated P_i and the demand D_i are random; the first can rely on the wind speed, solar irradiation intensity, etc.; and the second would be determined by uses of the energy on the part of the consumers. This gives rise to the surplus Q_i that will also be a random variable in the Smart Grid. Its value in a point in time will define an agent as a seller or an energy buyer [20].

A second hypothesis might bear in mind that the energy exchange will only happen among the Smart Grid (each other) or the substation. Then it won't be deemed the energy exchange with the macrogrid, which means that an electric possible transmission system will not be considered present [20].

All energy exchange that is carried out either among the Smart Grid and the Smart Grid and the substation incurs a cost associated with the energy losses in the driver. The energy losses in the feeders or the electric lines that are connected to each other, to the Smart Grid or to the substation are a function of the driver's resistance, the distance of the line, and of the power transmitted by the line in a specific time t.

2.1 Losses of power for the exchange between a Smart Grid and the substation

If a smart grid carries out an energy exchange with the substation, the losses of power incurred can be determined through Eq. (2) [20]:

$$P_i^{loss} = R_{io} I_o^2 + \beta P_i(Q_i) \tag{2}$$

where

P_i^{loss} is the losses due to the exchange of power between the substation and the Smart Grid $i \in N$.

R_{io} is the driver's resistance that joins the substation with the Smart Grid $i \in N$.

This resistance is calculated as the product of the resistivity per unit length of the driver in $[\Omega/\text{km}]$ used to connect both Smart Grid and the distance in $[\text{km}]$ between these elements.

I_o is the electric current in $[A]$ that flows through the driver, which joins the substation with the ith Smart Grid.

β is the coefficient that reflects the fraction of the losses in the transformer by the substation during the power exchange.

$P_i(Q_i)$ is the power flow between the substation and the i-th Smart Grid.

It can be said that the power losses associated to the power exchange are made up of a loss component in the electric line (feeder or sub-transmission line) that links the substation with the Smart Grid $i \in N$. The second component is given by the losses in the substation due to the use of the transformer to carry out the exchange of power. If it is considered that the electric current through the electric line of distribution may be calculated from:

$$I_o = \frac{P_i(Q_i)}{U_o}, \tag{3}$$

then, Eq. (2) to determine power losses can be written as an only power flow function through electric line given by Eq. (4):

$$P_i^{loss} = R_{io} \left[\frac{P_i(Q_i)}{U_o} \right]^2 + \beta P_i(Q_i) \tag{4}$$

The power flow depends on the kind of the Smart Grid $i \in N$ (buyer or seller); thus [1]:

$$P_i(Q_i) = \begin{cases} Q_i & if \quad Q_i > 0 \\ L_i & if \quad Q_i < 0 \\ 0 & other\ cases \end{cases} \tag{5}$$

Eq. (4) expresses the next; if a Smart Grid acts as a seller, the power of Q_i is completely sold to the substation; thus the flow power P_iQ_i corresponds exclusively to that power, and the power losses are determined by Eq. (4) [20].

On the other hand, if the i-th Smart Grid is a buyer, the power flow will be generated from the substation that should deliver such a power to solve the Smart Grid's power demands and its losses of power incurred for the power flow. Then the power of L_i which should be delivered by the substation is determined by [20]:

$$L_i = P_{io}^{loss} + P_i^{required} \tag{6}$$

where $P_i^{required} = -Q_i$ is the power required by the substation's load and the value of L_i the power flow through the line. When substituting these values from Eq. (4) to Eq. (6), the expression for the power which should be delivered by the substation to Smart Grid $i \in N$ is reached [1]:

$$L_i = R_{io}\left[\frac{L_i}{U_o}\right]^2 + \beta L_i - Q_i \tag{7}$$

$$\frac{R_{io}}{U_o^2} L_i^2 - (1-\beta)L_i - Q_i = 0 \tag{8}$$

Eq. (8) can present three possible solutions for the variable L_i because the same one corresponds to a quadratic equation.

If the equation presents real positive roots, the root that is the solution will be the lesser of the two, since it will cause fewer losses. Then the losses through the distribution line are determined to substitute in Eq. (4) the value of $P_iQ_i = |L_i|$.

If the equation presents negative roots or it does not have a real solution, the considered answer is:

$$L_i^* = \frac{(1-\beta)U_o^2}{2R_{io}} \tag{9}$$

Then the power losses are calculated substituting L_i^* in Eq. (7).

In either case, if N_b is the total number of buyers present in a certain time, being $N_b \subseteq N$, then it should be fulfilled with the power of the substation at a given moment that [1]:

$$\sum_{i \in N_b} L_i \le P_{subestation} \tag{10}$$

The value of L_i is the power flow which means the demanded power plus the power loss in the electric lines.

2.2 Power loss in smart grids

Suppose that the energy exchange is carried out between the Smart Grid $\boldsymbol{i} \in \boldsymbol{N_b}$ denominated buyer, and another Smart Grid $\boldsymbol{j} \in \boldsymbol{N_s}$ called the seller. Since $\boldsymbol{N_b}$, the group of all the Smart Grid buyers and the group of all the Smart Grid sellers with $\boldsymbol{N_b} \cup \boldsymbol{N_s} = \boldsymbol{N}$; the power losses will be similar to the case of the exchange with the substation, unless:

1. The energy exchange does not incur in the use of the transformer substation; consequently, the loss coefficient is $\beta = 0$.

2. The energy exchange between the Smart Grids should not necessarily be carried out with a voltage U_o but to a lower voltage U_1.

Thus, the energy losses for the power flow of a Smart Grid $i \in N_b$ y and another Smart Grid $j \in N_s$ can be determined from the equation [20]:

$$P_{ij}^{loss} = R_{ij} I_{ij}^2 \tag{11}$$

Since R_{ij} is the total resistance of the driver that joins the i-th Smart Grid buyer with the jth Smart Grid seller, their value is calculated from $R_{ij} = R \bullet d_{ij}$, something akin to an exchange case with the substation.

The current I_{ij} depends on the power flow through the electric line; therefore, as in the power exchange case with the substation and except for the different voltage level that is U_1, the power losses will be given for [20]:

$$P_{ij}^{loss} = R_{ij} \left[\frac{P_i(Q_i)}{U_1} \right]^2 \tag{12}$$

The power flow $P_i(Q_i)$ is similarly defined in Eq. (5) where the value of power L_i that supplies the Smart Grid buyer is determined by Eq. (11) with the value $\beta = 0$, getting the following equation [20]:

$$\frac{R_{ij}}{U_1^2} L_i^2 - L_i - Q_i = 0 \tag{13}$$

The solution of Eq. (13) will result once again in the cases that have been presented before where there are two different real solutions, a unique real solution, or no practical solutions. This way if:

Eq. (13) generates two values L_i, positive real and different. The lowest value is chosen since it will produce the fewest losses. The losses due to the power exchange are determined by (12) when substituting $P_i(Q_i) = |L_i|$.

On the other hand, Eq. (13) produces real roots, or there is no real solution; the value that is to be adopted for the power is:

$$L_i^* = \frac{U_1^2}{2R_{ij}} \tag{14}$$

The value L_i^* is replaced in Eq. (12) to determine the losses, since it is the power sum of the microgrid plus the power losses present during the flow power. The Smart Grid that acts at this precise point of time like seller will not necessarily cover the power Q_i required by the part of a certain Smart Grid buyer.

2.3 Algorithm for the coalition building in a cooperative game with transferable utility

To set an algorithm 1 based on [20], considerations and definitions may be carried out so that the result is a modified algorithm of [20] with the incorporation of restrictions and hypothesis that simplify the mathematical process and the calculations when carrying out its simulation.

As it was described in point (2.2), a group Smart Grid S of N is considered present in the electric grid linked to a macrogrid through a substation. Thus, a coalition game is formulated which is formed by the pair (N, v), since N is the total number of players (Smart Grid) and $v : 2^N \to \mathbb{R}$ is a function that assigns to a coalition $\subseteq N$ a real number to represent the total benefit reached by S. It must, therefore, define the value of function $v(s)$ assigned to this number or the coalition $S \subseteq N$.

The following describes the subroutines that would make up an algorithm, which will be necessary to set up the simulation that allows determining the game payment functions, the power loss of electrical grid, and power distribution in the cooperative exchange based on the resulting coalitions in the game process.

Algorithm 1 Coalition Forming Algorithm of Smart Grid

Initial State

Each coalition is one MG, which means that all MGs cannot form coalition with others. Therefore, the network is partitioned by $S = S_1, S_2, \cdots, S_N$.

Stage 1 Coalition Formation:

repeat

a) $M = Merge\ (S)$: The MGs will form coalition or merge small coalitions to big one

b) $S = Split\ (M)$: the MGs will decide to leave from coalitions to form new coalitions through the Pareto Order in (8)

until no MGs can do merge-and-split operation to get more payoffs, and the network is partitioned by S′.

Stage 2 Power transfer:

repeat for every $S_i \epsilon S'$.

The MGs $\in S_i$ will exchange power with others by the order of forming coalition.

until no local power transfer is possible.

If every $S_i \epsilon S'$, any seller or buyer, which has not meet its demand or has power surplus to sell, can exchange power with MS.

2.3.1 Subroutine for coalition formation

Once the noncooperative exchange is established, the next step is to form the coalitions which are the generation of cooperative groups to ease substation load and maximize the Smart Grid's profitability through the decrease of the losses [20].

Issues that should be considered by the time to begin to carry out the coalitions are the exchange between the Smart Grid regardless of the substation. Depending on the distance among Smart Grids in the subsets and at smaller distance minors, there will be losses; the exchange is carried out at the local level, without the necessity of the substation, except for it still existing as a surplus or lacking energy in the coalition and Smart Grid.

At the moment to start developing the coalitions, it is essential to consider that the exchange between the Smart Grid depends on the distance between the Smart Grid and the subsets S_b y S_s (at a shorter distance lower will be the losses). The exchange is carried out at the local level without the need of the substation except that an energy surplus or lack of power supply in the coalition and Smart Grid would be present.

The aim of forming coalitions inside the electric grid is to look for participation of group N, so the members of group N are creating disjoint subsets, where each subset $S_i \subset N$ is a coalition [21]. Thus, the participation established will be $\{S_1, S_2, S_3, \cdots, S_n\}$ [1]. As a coalition has a large number of possible combinations, it is necessary to introduce heuristic elements to simplify the calculations and reduce the operation number to calculate a conformed partition of a group of coalitions.

The first step is to determine the neighbors [22], defining like a neighboring coalition $S_i \subset N$ to that one with the shortest distance toward the other coalition $S_j \subset N$. In this point, the first restriction corresponding to the distance between coalitions appears. This distance is called threshold; d_{umbral} is the shortest distance that the two coalitions must have between them to be denominated neighbors, which correspond to the minimal losses of power that should be considered in such grid, so the energy quality indexes are inside the acceptable systems.

From this approach arises that large-size coalitions will hardly be formed; even a great coalition that involves all the members of the grid will be formed when the number of Smart Grid is significant.

Property: For the coalition game presented (N, v), the great coalition of all the Smart Grid rarely rises as the result of the presence of a series of expenses incurred by the power exchange, since the longer the distance, the bigger losses the grid will have. Rather, the disjoint independent coalition will be formed in the grid [22].

Observation: For the proposal of formation game coalitions (N, v), the size of any coalition $S_i \subset S$ that will be formed in the grid should satisfy the distance $d \leq d_{threshold}$.

The participation that will be carried out in the grid corresponds to merge the Smart Grid neighbors into a set of pairs in a way that each pair has a seller and a buyer that is located at the shortest reasonable distance and fulfill the distance restriction. In this first stage of coalition building, some Smart Grid can be initially isolated by the dynamics of the game. That means they do not fulfill with distance restriction, the number of Smart Grid is odd, the number of elements belonging to the set of the seller is greater or lesser than the number of the elements from the buyer set, and all the combinations are possible from these alternatives.

The next building coalition process follows the rule of coalition and division to achieve this; the following additional and necessary concepts are considered to understand the proposed algorithm.

Definition: Consider two sets of a disjoint independent coalition called $\mathcal{C} = \{C_1, C_2, C_3, \cdots, C_l\}$ and $\mathcal{K} = \{K_1, K_2, K_3, \cdots, K_m\}$ made by the same players (Smart Grid) that belong to the grid). Let $\phi_j(C_j)$ be the payment of player j in the coalition $C_j \epsilon \mathcal{C}$, and $\phi_j(K_j)$ the payment of player j in the coalition. Then, $\mathcal{C}$ is preferred for the collection $\mathcal{K}$ only if the Pareto principle is fulfilled that is shown by [1, 2]:

$$\mathcal{C} \triangleright \mathcal{K} \Longleftrightarrow \left\{ \phi_j(\mathcal{K}) \forall (\mathcal{K}) \forall j \in \mathcal{C}, \mathcal{K} \right\} \tag{15}$$

or at least just with a single player j that applies this expression.

Definition of the Pareto principle: The principle settles that the group of Smart Grid N prefers to be divided into partition or collection $\mathcal{C}$ rather than collection $\mathcal{K}$, if at least a player can improve his/her profitability when changing the structure of $\mathcal{K}$ to $\mathcal{C}$ without reducing the benefits or the payments of other players in the Grid [20]. To apply the Pareto principle, the process of coalition building will follow the coalition and division rules [23].

Definition of the coalition rule (merge): For a group of coalitions $\mathcal{S} = \{S_1, S_2, S_3, \cdots, S_n\}$, two or more coalitions decide to merge just if the profitability increases (it reduces the power losses) of at least a Smart Grid, without affecting or diminishing the profitability of the other members of the group N [23]:

$$\left\{U_{j=1}^{k} S_j\right\} \triangleright_m \{S_1, \cdots, S_k\} \tag{16}$$

Definition of the division rule (split): A coalition $\hat{S} = U_{j=1}^{k} S_j$ decides to be divided into two or more disjoint coalitions if just a Smart Grid increases its profitability (reduces the power losses) without affecting or diminishing the profitability of the other members of the group.

$$\{S_1, \cdots, S_k\} \triangleright_s \left\{U_{j=1}^{k} S_j\right\}$$

2.3.2 Subroutine for the exchange of power

As in the initial subroutine the group N of the Smart Grid was classified, these were split into buyers and sellers' subsets where the coalition is expressed as $\mathcal{S} = S_s \cup S_b$. However, it may focus on several approaches to the distribution of energy for the assignment of the sellers to the buyers. The approach outlined is the preference of the buyers in the coalition.

The split $\mathcal{S}$ with k buyers in $S_b \cup \mathcal{S}$, being $S_b = \{b_1, b_2, b_3, \cdots, b_k\}$, and buyers in $S_S \subset$, being $S_S = \{s_1, s_2, s_3, \cdots, s_S\}$, these groups will act sequentially. An important consideration is the local transfer of energy made by the seller and buyer before using the substation.

Also, if a Smart Grid just buys or sells energy from or toward the substation, this Smart Grid is left out of the Grid N since it does not deliver any benefit to the coalition.

3. Simulation of the electric system based on the theory of games

The software Matlab 2017 and the data of the network of **Figure** 1 were used for the simulation.

3.1 Input data

Table 1 shows the data entered in the simulator; they include the driver resistance, link voltage, and the minimum threshold distance to build coalitions.

Table 2 shows the substation characteristics, such as, geographical location, power, meter of energy losses for the transformer of the substation, and price of the electricity in dollars per MW.

In **Table 3**, the data of 10 microgrids (MG) [24] that are composed of 6 buyers (−1) and 4 sellers (+1) are shown. Additionally, the location is given in Km by a Cartesian coordinate system, power generated by the MG, energy demand by each MG, and the price of electricity.

3.2 Analysis of the results

3.2.1 Noncooperative model

Table 4 shows the algorithm results for the noncooperative model. The energy surplus is higher than zero (Q_i>0), in which the Smart Grid has an energy surplus

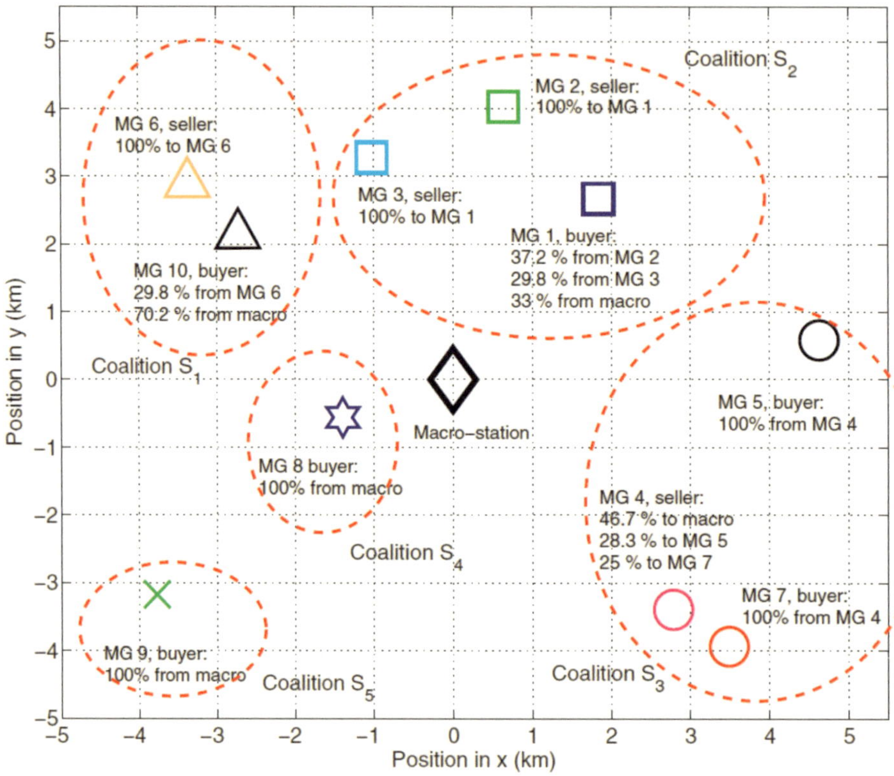

Figure 1.
Base model.

Resistance [Ω/km]	MT voltage [kV]	BT voltage [kV]	Threshold distance [km]
R	Uo	U1	Du
0.0147523	22	22	5

Table 1.
Data feeder.

N°	Location [km]		Power [MW]	Loss constant	Cost of energy [$/MW]
	X	Y			
0	0	0	100	0.02	1

Table 2.
Macro station or substation data.

so that it can sell that power (seller). Likewise, it is observed that there are other values of de $Q_i < 0$; consequently, in this case, some MGs need to buy energy from another MG or directly from the substation. The value of L_i is the power flow, that is, the demanded power plus the power loss in the electric lines. Finally, there are values of the power losses P_{i0}, and the individual payments P_{i0}.

3.2.2 Coalition building

When Smart Grids decide to build coalitions with its neighbors, the merger processes and the application Pareto principle generate a stable coalition where the

N°	Location [km]		Power [MW]	Demand [MW]	Energy price [$/MW]	State buyer: (−1) seller: (1)
	X	Y				
1	1.8	2.6	0	152.2	1	−1
2	1.6	4	56.6	0	1	1
3	−1	3	45.4	0	1	1
4	2.8	−3.3	134.3	0	1	1
5	4.7	0.4	0	35.4	1	−1
6	−3.4	2.8	42	0	1	1
7	3.5	−4	0	33.2	1	−1
8	−1.4	−0.6	0	60	1	−1
9	−3.8	−3.2	0	68	1	−1
10	−2.8	2.2	0	140.9	1	−1

Table 3.
Smart grid data.

N°	Q_i [MW]	L_{i_optimo} [MW]	P_{io} [MW]	U_{ii}
1	−152.2000	162.0883	9.8883	−9.8883
2	56.6000	54.5463	2.1686	−2.1686
3	45.4000	44.0290	1.4294	−1.4294
4	134.3000	126.2559	8.9307	−8.9307
5	−35.4000	36.6394	1.2394	−1.2394
6	42.0000	40.6068	1.4616	−1.4616
7	−33.2000	34.3907	1.1907	−1.1907
8	−60.0000	61.6978	1.6978	−1.6978
9	−68.0000	71.4586	3.4586	−3.4586
10	−140.9000	150.3462	9.4462	−9.4462
Average of power losses for the noncooperative case is 4091 [MW].				

Table 4.
Noncooperative state.

members of each coalition can improve their payments. The evolution of the payments can also be observed and compared with the case presented in [24]. The payments are shown in **Table 5**. In analyzing the payments concerning pattern [24], these improve when the Smart Grids decide to build coalitions like those shown in **Figure 2**.

It is noteworthy that like [24], it was not possible to improve the payment of the Smart Grid 9, which was left isolated for the cooperative game and Pareto principle. It would not represent any problem since it does not contribute any benefit to the coalition's members nor does it worsen the payments.

3.2.3 Power exchange in a cooperative game

Power exchange in a cooperative game incorporates the restrictions in the coalition building, improving the algorithm presented for [24]. It carries an

−9.8883	−5.7107
−2.1686	−1.2524
−1.4294	−1.4254
−8.9307	−5.2128
−1.2394	−1.2394
−1.4616	−0.7797
−1.1907	−0.6950
−1.6978	−1.6930
−3.4586	−3.4586
−9.4462	−5.0395

Table 5.
Vector payment evolution.

Figure 2.
The coalition formed for the cooperative game in the energy exchange.

improvement of the reduced power losses. **Table 6** shows the increases in the payments of the members belonging to the grid.

3.2.4 Energy exchange in the grid

Finally, **Table 7** shows the power bought to each MG and substation during the process of energy exchange in the grid.

Seller i	Buyer j	Transferred power [MW] $MG_i - MG_j$ P_{ij}	Power purchased from the substation [MW] P_{j0}	Power sold to the substation [MW] P_{i0}
		Coalition $S_1\{1,2\}$		
2	1	2.2507	—	—
Substation	1	—	4.7124	—
		Coalition $S_2\{5\}$		
Substation	5	—	—	1.2394
		Coalition $S_3\{4,7\}$		
4	7	0.4390	—	—
4	Substation	—	—	5.4688
		Coalition $S_4\{3,8\}$		
3	8	2.7257	—	—
Substation	8	—	0.3926	—
		Coalition $S_5\{9\}$		
Substation	9	—	—	3.4586
		Coalition $S_6\{6,10\}$		
6	10	0.6009	—	—
Substation	10	—	5.2183	—

Table 6.
Power exchange in the distribution grid.

	Purchase to	MW	Losses	Transferred power
		Coalition $S_1\{1,2\}$		
MG_1	MG_2	164.339	2.2507	162.0883
MG_1	Substation	61.3124	4.7124	56.6000
		Coalition $S_2\{5\}$		
MG_5	Substation	46.6394	1.2394	45.4000
		Coalition $S_3\{4,7\}$		
MG_7	MG_4	134.739	0.4390	134.3000
MG_4	Substation	42.1082	5.4688	36.6394
		Coalition $S_4\{3,8\}$		
MG_8	MG_3	44.7257	2.7257	42.0000
MG_8	Substation	34.7833	0.3926	34.3907
		Coalition $S_5\{9\}$		
MG_9	Substation	65.1564	3.4586	61.6978
		Coalition $S_6\{6,10\}$		
MG_{10}	MG_6	72.0595	0.6009	71.4586
MG_{10}	Substation	155.5645	5.2183	150.3462

Table 7.
Energy exchange in the distribution grid.

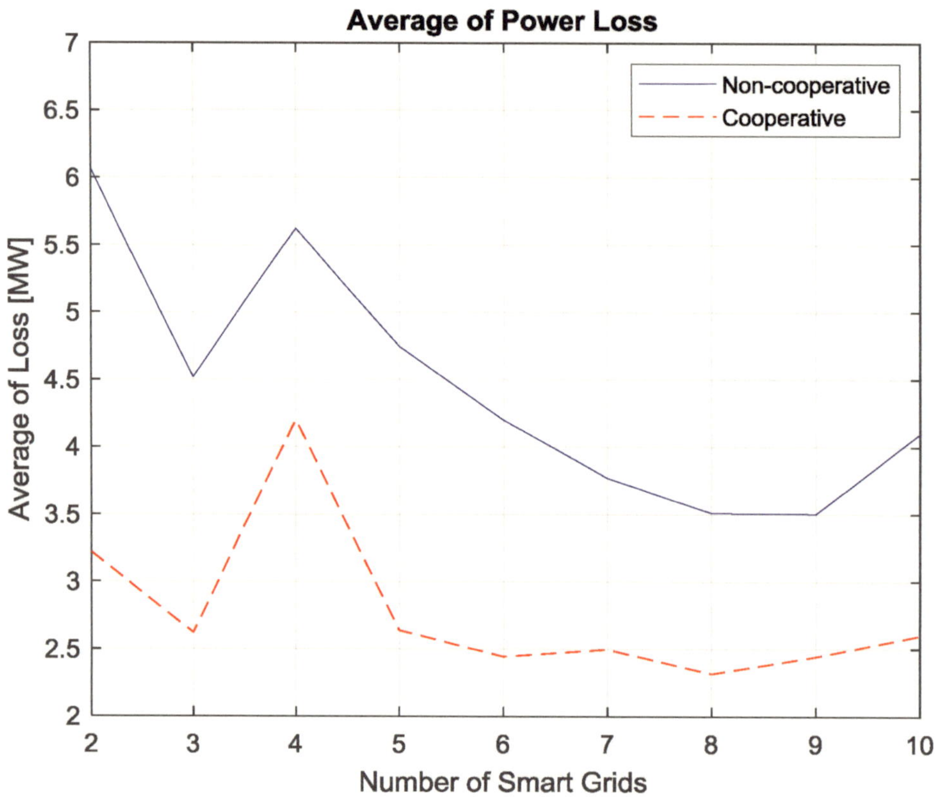

Figure 3.
Average power losses in a cooperative system vs. noncooperative.

3.2.5 Average loss

Figure 3 shows that as *N increases*, the power losses tend to reduce. When *N* is big in a smart grid, it has higher possibilities to find neighboring nodes to develop the coalition process for cooperative exchange of energy.

Table 8 shows a significant power loss in the cooperative model compared to the noncooperative. Thus, the study concludes that the average losses decrease by 38.62 when they join the MGs.

N° MG	Noncooperative	Cooperative	[%]
2	6.0622	3.2160	46.95
3	4.5179	2.6205	42.00
4	5.6211	3.6968	34.23
5	4.7448	2.6382	44.40
6	4.1976	2.4421	41.82
7	3.7680	2.4953	33.78
8	3.5092	2.3150	34.03
9	3.5036	2.4420	30.30
10	4.0979	2.4542	40.11

Table 8.
Payment evolution.

4. Conclusions

The most outstanding conclusion in this chapter is the development of a coalition building algorithm through the game theory to reduce energy losses in smart grids, which are based on a conceptual new model within the same ones that concentrate on the consumers and benefits if they decide to use the flexibility of distributed generation grids.

The coalitions built between MGs could be very profitable if they were truly allowed, and it they will encourage the consumers to participate and to take the next step as prosumers, that is, to produce and consume energy at the same time.

The proposal presented allows the MG building coalitions to minimize the power loss when the power is transmitted from an MG to another MG, to the macro station, or to the nearest substation.

This study simplifies numeric calculations, by introducing certain heuristics to the algorithm, through the approximation of the data that belongs to an ideal or practical system. That is, a great coalition. among all the participants is not possible.

It can be seen that for similar distances between a buyer and a seller, and a buyer and the substation, the power losses can end up being lower in the second case than the first. This is because it is in the voltage level between the interconnection, which is lower when two MGs are connected, instead that an MG and the substation: $U\_1 < U\_2$.

Concerning the theoretical pattern, the losses significantly decrease by introducing into the coalition building the right restrictions such as the correct selection of neighbors (threshold distances), load priorities (distribution of power in the coalitions), power flow, and limitation of the energy in Smart Grid.

About the theoretical pattern, the losses significantly decrease by introducing appropriate restrictions into the coalition building, such as the correct selection of neighbors (threshold distances), load priorities (distribution of power in the coalitions), power flow, and limitation of the energy in Smart Grid.

Author details

Javier B. Cabrera[1,2], Manuel F. Veiga[2], Diego X. Morales[1]* and Ricardo Medina[3]

1 Smart Grid Research Group, Catholic University of Cuenca, Cuenca, Ecuador

2 AtlanTTIC, Vigo University, Vigo, Spain

3 Luis Rogerio Gonzalez Institute—Senescyt, Cuenca, Ecuador

*Address all correspondence to: dmoralesj@ucacue.edu.ec

References

[1] Rheinisch-Westfä lische Elektrizitä tswerke (RWE). Typical daily consumption of electrical power in Germany. 2005

[2] Arefifar SA, Mohamed YAI, El-Fouly THM. Supply-adequacy-based optimal construction of microgrids in smart distribution systems. IEEE Transactions on Smart Grid. 2012;**3**(3):1491-1502

[3] Niyato D, Wang P. Cooperative transmission for meter data collection in smart grid. IEEE Communications Magazine. 2012;**50**(4):90-97

[4] Ochoa LF, Harrison GP. Minimizing energy losses: Optimal accommodation and smart operation of renewable distributed generation. IEEE Transactions on Power Systems. 2011; **26**(1):198-205

[5] Kantarci B, Mouftah HT. Cost-aware smart microgrid network design for a sustainable smart grid. In: Proc. GC Wkshps; 2011. pp. 1178-1182

[6] Meliopoulos S, Cokkinides G, Huang R, Farantatos E, Choi S, Lee Y, et al. Smart grid infrastructure for distribution systems and applications. In: Proc. 44th HICSS; 2011. pp. 1-11

[7] Deilami S, Masoum AS, Moses PS, Masoum MAS. Real-time coordination of plug-in electric vehicle charging in smart grids to minimize power losses and improve voltage profile. IEEE Transactions on Smart Grid. 2011;**2**(3): 456-467

[8] Vargas A, Samper ME. Real-time monitoring and economic dispatch of smart distribution grids: High performance algorithms for DMS applications. IEEE Transactions on Smart Grid. 2012;**3**(2):866-877

[9] Costabeber A, Erseghe T, Tenti P, Tomasin S, Mattavelli P. Optimization of micro-grid operation by dynamic grid mapping and token ring control. In: Proc. 14th EPE; 2011. pp. 1-10

[10] Costabeber A, Tenti P, Mattavelli P. A surround control of distributed energy resources in micro-grids. In: Proc. IEEE ICSET; 2010. pp. 1-6

[11] Tenti P, Costabeber A, Trombetti D, Mattavelli P. Plug and play operation of distributed energy resources in micro-grids. In: Proc. 32nd INTELEC; 2010. pp. 1-6

[12] Kirthiga MV, Daniel SA, Gurunathan S. A methodology for transforming an existing distribution network into a sustainable autonomous micro-grid. IEEE Transactions on Sustainable Energy. 2013;**4**(1):31-41

[13] Li Z, Wu C, Chen J, Shi Y, Xiong J, Wang AY. Power distribution network reconfiguration for bounded transient power loss. In: Proc. IEEE ISGT Asia; 2012. pp. 1-5

[14] Bouhouras AS, Andreou GT, Labridis DP, Bakirtzis AG. Selective automation upgrade in distribution networks towards a smarter grid. IEEE Transactions on Smart Grid. 2010;**1**(3): 278-285

[15] Katic NA, Marijanovic V, Stefani I. Profitability of smart grid solution application in distribution network. In: Proc. 7th MedPower; 2010. pp. 1-6

[16] Fadlullah ZM, Nozaki Y, Takeuchi A, Kato N. A survey of game theoretic approaches in smart grid. In: Proc. IEEE WCSP; 2011. pp. 1-4

[17] Couillet R, Perlaza SM, Tembine H, Debbah M. A mean field game analysis of electric vehicles in the smart grid. In: Proc. IEEE INFOCOM Workshops; 2012. pp. 79-84

[18] Zhang C, Wu W, Huang H, Yu H. Fair energy resource allocation by minority game algorithm for smart buildings. In: Proc. DATE; 2012. pp. 63-68

[19] Saad W, Han Z, Poor HV. Coalitional game theory for cooperative micro-grid distribution networks. In: Proc. ICC; 2011. pp. 1-5

[20] Saad W, Han Z, Vicent PH. Coalitional game theory for cooperative micro grid distribution networks. In: IEEE International Conference on Communications Workshops (ICC); 2011

[21] Lena M. A merge and split mechanism for dynamic virtual organization formation in grid. IEEE Transactions on Parallel and Distributed Systems. 2014;**25**(3):3

[22] Saad W, Han Z, Poor HV. Coalitional game theory for cooperative micro-grid distribution networks. In: IEEE Xplore; 2011. pp. 1-5

[23] Apt K, Witzel A. A generic approach to coalition formation. International Game Theory Review. 2009;**11**(3): 347-367

[24] Magaña Nieto A. TDX, 30 10 2015 [En línea]. Available from: http://www.tdx.cat/TDX-0722109-095713

Section 2

Communication Standards and Solutions

Chapter 4

ICT Technologies, Standards and Protocols for Active Distribution Network Automation and Management

Mohd Asim Aftab, S.M. Suhail Hussain and Ikbal Ali

Abstract

The concept of active distribution network (ADN) is evolved to address the high penetration of renewables in the distribution network. To leverage the benefits of ADN, effective communication and information technology is required. Various communication standards to facilitate standard-based communication in distribution network have been proposed in literature. This chapter presents various communication standards and technologies that can be employed in ADN. Among various communication standards, IEC 61850 standard has emerged as the de facto standard for power utility automation. IEC 61850-based information modeling for ADN entities has also been presented in this chapter. To evaluate the performance of ADN communication architecture, performance metrics and performance evaluation tools have also been presented in this chapter.

Keywords: active distribution networks (ADNs), communication standards, communication technologies, co-simulation, IEC 61850, network latency

1. Introduction

With increasing impetus towards use of cleaner energy, power utilities are increasingly integrating renewable energy resources (RER) at distribution level. High penetration of RERs and distributed energy resources (DERs) in the distribution network poses the challenge of reliable operation, control and quality power supply. The concept of active distribution network (ADN) will allow distribution networks to integrate DERs efficiently by addressing the above challenges by incorporating Information and Communication Technologies in distribution systems [1]. The management of ADN with penetration of DERs and RERs, which are highly intermittent, can be achieved by coordinated operation of different components of ADN such as distribution system operator (DSO), control centers, DERs, distribution substations and other components. In ADNs the different components are geographically distantly placed, thus for coordinated operation a wide area communication infrastructure is required. For stable, reliable and efficient management of distribution network, this communication is required to be standardized, interoperable, securable and scalable [2].

To facilitate standard communication in the power distribution networks, various standards and protocols such as IEEE P2030, IEC 60870-5, IEEE C37.118.1/2, Distributed Network Protocol (DNP3), IEC 61850, IEC 61970 and IEC 61968, OpenADR, etc., have been developed. This chapter will present a comprehensive analysis of these state-of-the-art standards and protocols for application in ADNs. The challenges presented by these standards are of feasibility, flexibility and interoperability. In this regard IEC 61850 is emerging as one of the most popular and widely accepted solutions since it is based on the interoperability approach and provides flexibility in implementation [2].

For information exchange between different components of ADN different communication technologies have been explored in literature [3, 4]. This chapter enlists and provides a detailed study of different communication technologies that can be employed in ADNs. The presence of RERs and DERs in ADNs introduces intermittency, thus the communication architecture for ADN must be highly scalable. To address the scalability issue in the smart grid communication architectures Web protocols have been employed. The proposed chapter will present a detail study of the different Web protocols and their suitability for ADNs [5].

The performance of different communication network architectures is evaluated for network latency, quality-of-service (QoS), robustness, reliability and data security to determine its applicability and suitability in ADNs. This chapter will discuss the different simulator tools for evaluating the performance of ADN communication networks. An overview of different real-time test beds and state-of-the-art co-simulation platforms of interfacing power system simulators and ICT simulators will be presented.

2. Active distribution networks: the concept

With the integration of large amount of renewable energy sources into the distribution network, the current distribution network has evolved into an active network from a passive network. The high penetration of DGs into the distribution system introduces bidirectional power flows in the distribution network and also causes voltage rise and increased levels of fault currents. The major challenge is to monitor, control and manage the ADN in order to supply reliable and clean energy to the consumer.

2.1 Challenges with traditional distribution system

In traditional distribution system, the DSOs operated in a top down approach in which the electricity from the transmission system operator (TSO) was received at the DSO level and was transferred to distribution network operator and finally to the end consumer. Since the distribution system operated as a radial network with the source supplying the loads at the end consumer in a unidirectional fashion, there were predictable electricity flows in the network which does not require extensive management and control.

However, with the proliferation of DGs in the distribution network introduces new challenges in management and control of the network and in ensuring reliable power and quality power supply to the consumer. Also increasing number of these DGs are being connected to the distribution network, both in capacity and numbers, leads to unpredictable power flows in the network, wide voltage variations, changed network reactive power characteristics and increased levels of fault current. This has resulted in profound implications on the operation of DSOs.

Thus, the DSOs are expected to operate the modified distribution network (i.e., ADN) in a more secure and reliable way and provide high service quality to the end consumers.

2.2 Evolution of active distribution network

The growing impetus towards renewable energy sources (RESs) like wind power plants, photovoltaic power (PV) plants, fuel cells, electric vehicles (EV) supporting vehicle to grid (V2G), combined heat and power (CHP) plant, etc., is expected to increase their share in total power generation capacity worldwide. This has evolved the concept of Distributed Generation (DG) which involves small or medium generating units usually located on-site. These DGs meet the local power needs and dispatch the remaining power to the electric grid. In a move to reduce carbon footprints and increase supply flexibility and reliability these DGs are getting largely integrated into the distribution system and changing its very nature from passive to active which evolves the concept of active distribution network (ADN).

In order to make the most efficient use of the existing network infrastructure and manage DGs for reliable and secure power supply, the concept of ADN management is evolved. ADN will allow efficient integration of DGs to the existing distribution infrastructure by taking maximum advantage of the inherent characteristics of DGs. This requires the planning and operation of the distribution network by taking into account the bidirectional power flows in the network. The system planning and development could happen only by setting out implications for the DSOs, TSOs and for the DG owner/operator.

In an ADN, the DGs provide advantages such as security of power supply, loss reduction in transmission and distribution, peak load and congestion reduction and less network investments. However, most of the DGs are nondispatchable in nature and therefore matching of power production profile with that of the load demand profile cannot be guaranteed at all times. It might happen that the DGs do not generate enough power in cases when the distribution network is constrained while DGs provide abundance power supply when the demand is low. Thus, there remains a challenge in operation of ADN and requires adequate mechanisms to provide solutions to these problems.

2.3 Architecture of active distribution network

The DGs are usually managed and controlled at the individual level and are geographically distributed in nature. The power requirements of the electrical grid are shared with the DG owner via aggregator. Since, management and control of individual DG is not possible due to large number of devices, the concept of aggregator plays a crucial role. Also, since the small scale DGs does not fall under the direct supervision and control of DSOs and thus DSOs acquire services from aggregators for monitoring and control of the DGs. The aggregator manages and controls a group of closely located DGs. The combined demand and supply from the group of DGs is shared with the Distribution System Operator (DSO) via the aggregators. The DSO develops the dispatch schedule for the DGs and provides it to DG owners via aggregators. **Figure 1** illustrates a schematic of ADN architecture including the aggregators and DSO.

The DG operator is in direct contact with the aggregator for providing dispatch schedules. Due to heterogenous nature of DGs in terms of capacity and density, the schedules for different DG owners are different which depends upon the connected sources to the system and the load density. The advantages for having a hierarchical architecture for ADN are as follows:

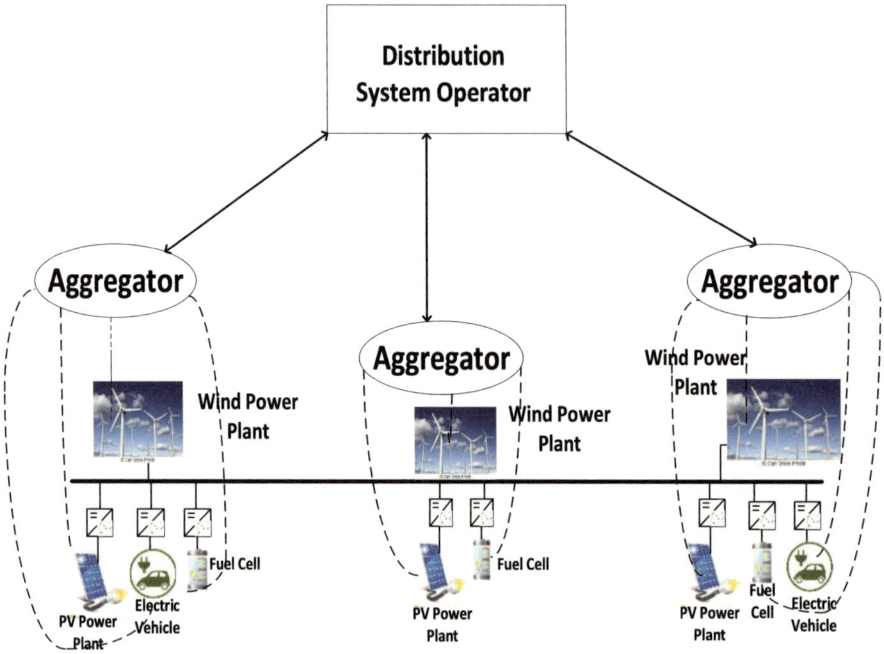

Figure 1.
A simplified schematic illustration of active distribution network hierarchy.

1. Real time information exchange between the TSO, DSO, Aggregator and DG owner.

2. DGs connected through aggregators would provide various services such as power reserves, regulating/balancing power, reactive power support, etc.

3. Easy scheduling between various actors/components of the ADN.

4. Congestion management of the distribution network by active control of DGs.

3. Communication configurations for ADNs: standards and technologies

The cornerstone for the ADN management is the ability of multiple entities such as DG operators, aggregators, DSOs and TSOs to interact with each other for providing monitoring, control and real-time exchange of energy consumption and power usage data. Also, the TSOs and DSOs can retrieve consumer usage data and online pricing and optimize the electricity distribution based on the electricity consumption via a communication network. Thus, a fast, reliable and secure communication infrastructure plays a vital role in management of ADN. With the adoption of information and communication technologies (ICT) in the ADN management it is possible to enhance the efficiency of power generation, transmission and distribution. The ICT helps in accumulating information from every point of ADN and can be used for demand forecasting, network planning, ADN operation, Control and Protection of ADN and in optimizing performance of ADN.

Due to increasing level of DG penetration, real time information about various measurements such as currents, voltage, active power, reactive power, etc., is becoming necessary. These measurements are required to be exchanged continuously between various components of ADN for overall monitoring and control applications. Also, due to intermittent nature of renewable DGs, these measurements are varying in relatively fast fashion such as like of meteorological data. Thus, a fast and reliable ICT infrastructure is required in order to improve the observability of the whole ADN using real time monitoring of the network. With evolving ICT infrastructure, the future ADN monitoring might not be limited to typical power system data monitoring rather than monitoring of new parameters such as dynamic phasor, dynamic line rating, rate of change of frequency, etc. These new measurements may be then utilized in disturbance management, predictive maintenance and to enhance the stability and load ability of the power system.

The components of ADN are distributed over vast geographic areas and thus in order to have a coordinated operation it requires a wide area communication infrastructure. The wide area network forms a backbone in providing communication between various ADN components to provide a reliable, secure, expedient and trustable service.

3.1 Communication standards for ADN

3.1.1 IEEE 1547

The IEEE 1547 [6] standard provides rules for interconnecting various distributed resources (DRs) to the electrical power systems (EPS). It is characterized by various forms of DRs operations and their interconnecting issues. It sets forth the guidelines for DG participation for voltage regulation, active power management, grounding requirements and integration of DR islands with the existing power systems. The guidelines for monitoring, control and information exchange among the DG and EPS is also provided by the standard.

The IEEE 1547 standard series has various parts dedicated to the issues related to DR interconnection to EPS. These are as follows:

1. IEEE 1547.1: IEEE Standard Conformance Test Procedures for Equipment Interconnecting Distributed Resources with Electric Power Systems.

2. IEEE 1547.2: IEEE Application Guide for IEEE 1547, IEEE Standard for Interconnecting Distributed Resources with Electric Power Systems.

3. IEEE 1547.3: IEEE Guide for Monitoring, Information Exchange, and Control of Distributed Resources Interconnected with Electric Power Systems.

4. IEEE 1547.4: Draft Guide for Design, Operation, and Integration of Distributed Resource Island Systems with Electric Power Systems.

5. IEEE 1547.5: Draft Technical Guidelines for Interconnection of Electric Power Sources >10 MVA to the Power Transmission Grid.

6. IEEE 1547.6: Draft Recommended Practice for Interconnecting Distributed Resources with Electric Power Systems Distribution Secondary Networks.

3.1.2 IEEE P2030

IEEE P2030 [7] standard is the first standard drafted by IEEE for providing smart grid interoperability. It provides a roadmap for establishing the framework for developing IEEE national and international body of standard aimed at development of a standard for smart grid which merges the disciplines in power applications, information technology and control through communications. The IEEE P2030 standard establishes the smart grid interoperability reference model (SGIRM) which develops a base terminology for providing functional performance, characteristics, engineering principle evaluation related to the smart grid interoperability. The SGIRM approach consists of systems of systems and inherently allows for extensibility, scalability, and upgradeability. The SGIRM approach is based on integration of power systems, communication and information technology. Also, it defines tables and data classification flow which are necessary for providing smart grid interoperability. According to IEEE P2030 standard, interoperability is defined as capability of a network, system, device to seamlessly transfer and exchange information with its counterpart in a secure and effective way.

The term Smart Grid interoperability is defined as the ability to effectively communicate and transfer information seamlessly among various devices, organizations even if they are using different variety of infrastructure and are spread along different geographic regions and locations. The smart grid interoperability is associated with three components: Hardware/software component, data formats, interoperability on content level. At the hardware/software level, the interoperability is achieved by developing or designing the devices which follow a standard blueprint and adheres to a common protocol. At the data format level, the messages or information must be encoded in a standard well defined syntax. At the content level, a common understanding of the meaning of the data/content being exchanged must be developed to achieve interoperability at content level. To transform the legacy networks into intelligent devices which can participate in smart grid communication, the standard must address the requirements of stakeholders and develop interoperable solutions and flexible business processes.

3.1.3 IEC 60870-5

IEC 60870-5 [8] standard was developed by IEC Technical Committee 57 to provide protocol for sending basic telecontrol messages from the telecontrol master station to outside stations which are connected through some form of permanent communication link. The telecontrol messages are transferred between the telecontrol equipment in the form of coded serial data which is used for monitoring and controlling of wide are processes. The part 5 of 60870 defines the interoperability among the telecontrol equipment. This standard is a combination of application layer of IEC 60870-5-101 and transport layer of TCP/IP standard. Within the TCP/IP, there is an independent choice of telecommunication networks such as X.25, ATM, and Frame relay.

The IEC 60870-5 supports unbalance and balanced mode of data transfer, provides unique addresses for master telecontrol stations, time synchronization facility, data classification facility and cyclic data updating facility.

3.1.4 IEEE C37.118.1/2

The IEEE C37.118 is the standard drafted by IEEE for synchronized phasor measurement in power system. It is the main standard which governs phasor measurement unit (PMU) operation. A PMU is a device which provides accurate time stamping of power system information by performing synchrophasors measurements by incorporating GPS time signal for time reference. It transmits synchrophasors data to remote peers either by unicast or multicast [9]. The IEEE C37.118 standard is split into two parts viz. IEEE C37.118.1 for synchrophasors measurement and IEEE

C37.118.2 for synchrophasors communication. Both these parts of IEEE C37.118 standard form the backbone of PMU operation and communication in power system.

The IEEE C37.118.1 [10] defines synchrophasors, frequency and rate of change of frequency (ROCOF) measurements. IEEE C37.118.1-based measurements made at various locations by the PMU can be readily obtained and interpreted at the Phasor Data Concentrator (PDC) accurately due to the presence of GPS time tagging on PMU information. The IEEE C37.118.1 does not specify underlying hardware or components required for carrying out such synchrophasors measurement. The IEEE C37.118.1 standard specifies certain synchrophasors measurement requirements which are as follows: synchrophasor estimation, frequency and ROCOF estimation, measurement reporting delay, measurement response time and measurement errors.

The PMU must measure the synchrophasor data according to the synchrophasor measurement and estimation as specified in the standard. Measurement latency is the time delay occurred from the instance an event occurs in the power system to time it is reported. Measurement response time is the time transition between the two steady state measurements when an input signal is applied. The purpose of having measurement response time is to ensure that the time tagging is working correctly in the PMU data. The measurement errors are usually computed as the total vector error (TVE) in the synchrophasor measurement by the PMU.

The IEEE C37.118.2 [11] standard defines a method of exchange of synchrophasor data between the power system devices. It provides the guidelines for data message formats which are to be exchanged between a PMU and PDC. It defines various messages which are exchanged for realizing a handshake operation between the PMU and PDC. The following type of messages are employed in synchrophasor measurement viz, data, configuration, header and command.

3.1.5 Distributed network protocol (DNP3)

Distributed network protocol (DNP3) [12] was drafted for providing open, interoperable communication among substation computers, IEDs, remote terminal unit (RTUs) and master stations in the electric utility industry. DNP3 was developed by the combined efforts of IEC TC 57 working group (WG-3) who have been working on OSI three layer “enhanced performance architecture (EPA)” for telecontrol applications. DNP3 is also the recommended practice for RTU to IED communication protocol.

DNP3 was first developed by Harris, Distributed Automation Products (originally Westronic, Inc.) and later it is managed by the DNP3 users group which is composed of vendors and electric utilities which are using the DNP3 protocol. Amendments and modifications in the current draft of DNP3 are carried out by the DNP3 users technical group. To ensure interoperability, longevity and upgradeability of DNP3 protocol, the modifications and recommendations are made open to DNP3 technical group. DNP3 is not limited to serial communication inside the substations but the widespread functionality of DNP3 make it usable with TCP/IP networks having Ethernet, frame relay, fiber-optic-based communication media.

3.1.6 IEC 61850

IEC 61850 has emerged as the global standard for substation automation system since its publication in the year 2004 [3]. The IEC 61850 standard is intended to provide interoperability among substation. The IEC 61850 standard was initially drafted for substation automation system and later on it was expanded to cover power utility system. IEC 61850 adopts object oriented approach for modeling power system components. Due to its worldwide acceptance by the industry and research organizations, it is poised to be the future automation industry standard.

Communication is divided into four main parts viz. Information modeling, services modeling, communication protocols and telecommunication media. Information modeling deals with the type of data that is to be exchanged. It is synonymous with noun in English language. Service modeling deals with reading, writing or other actions taken on data and is analogous to verb. Communication protocol is a way of mapping the data to the required action. Telecommunication media is the physical medium used for data communication.

IEC 61850 models power system components in terms of logical nodes and data objects. This modeling is known as information modeling. Information modeling is a way of exchanging standardized information as per the standard. The group of data objects that serves specific function is known as logical nodes and a group of logical nodes forms a logical device. Logical nodes may reside in different devices and at different levels. The objective of the standard is to specify requirements and to provide a framework to achieve interoperability between the IEDs supplied from different suppliers.

Based upon application, IEC 61850 defines main types of communication services viz. services for real time communication, services for client server communication and services for time synchronization. Services for real time communication are generic object oriented substation event (GOOSE) and sampled values (SVs). Due the time criticality of GOOSE and SV messages, they are mapped directly onto the Ethernet layer of OSI seven layer communication model. The standard specifies the protocol data unit for the GOOSE and SV messages. The SV are multicast in the network.

Whenever a fault occurs, protection devices respond to the fault by generating burst of GOOSE messages. The occurrence of fault changes the periodic heartbeat nature of GOOSE message into burst mode. In burst mode, the transmission interval of GOOSE increases sequentially such that after the certain time of trigger of the event, the retransmission time changes back to normal periodic nature as shown in **Figure 2**. As an event occurs (such as a fault) the retransmission time of GOOSE message is changed from To to T1, T2, T3, ..., Tn such that T1 < T2 < T3 < .... < Tn. The sequential increase in retransmission time ends until Tn reaches to To. The gradual increase in retransmission time in bursts is adopted in order to increase reliability of the network, since the GOOSE message conveys critical commands.

The GOOSE messages are LAN-based messages having no Internet or IP layer and is intended for protection purposes. In order to transport GOOSE messages over WAN, tunneling has been employed [13]. Also, differential protection in substations using IEC 61850 has been presented in [14]. Due to presence of distributed generations in a microgrid, the fault current levels increase and a revamped protection strategy is required. This protection scheme must me communication based so that relays are made aware of any addition or deletion of distributed generation. In [15, 16], authors have proposed microgrid protection strategy based on IEC 61850 communication.

The IEC 61850 standard specifies set of abstract services and objects that allows applications to work in a manner independent of the underlying protocol. These services are followed by vendors for invoking any functionalities and are known as abstract communication service interface (ACSI).

3.2 IEC 61850-based communication configuration for ADN

For designing IEC 61850-based communication architecture for ADN, the entities or components of the ADN are modeled as per the IEC 61850 standard.

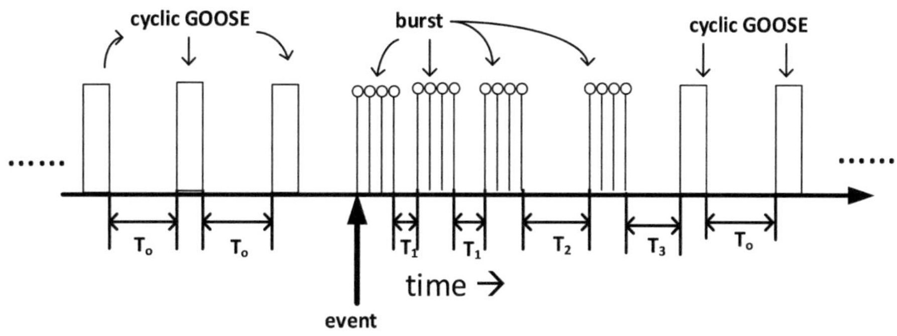

Figure 2.
GOOSE message retransmission in IEC 61850.

There are basically three types of IEDs in a distribution system viz. merging unit (MU) IED, Breaker IED and protection and control (P&C) IED. MU is the main equipment in process level, which receives current and voltage samples from non-conventional instrument transformers and then convert them to digital data packets and communicate to other IEDs, as per communication mechanisms described in IEC 61850-9-2LE [17]. The SV data generated from MUs is time stamped and synchronized using time synchronization source in the substation. A synchronizing accuracy of 1 μs is required by the "IEC 61850-9-2LE" process-bus implementation guidelines to synchronize the MUs in SAS. Breaker IED represents the circuit breaker controlling device, which controls and monitors the status and condition of breaker and also acts as a sink for tripping, close and interlocking commands. P and C IEDs normally receive the SVs data packets from MU IEDs and implement protection and control functions by exchanging appropriate data with other IEDs.

3.2.1 IEC 61850 information models for different components of distribution networks

To enable IEC 61850-based approach for ADN, information models for various entities of ADN are required to be modeled as per IEC 61850 standard. This modeling requires realizing ADN components in terms of logical nodes and data objects. Modeling of various power system components by using relevant logical nodes as per IEC 61850 has been proposed in the standard. The following parts of the IEC 61850 standard are for modeling different components such as,

- IEC 61850-7-420: Information modeling of various DERs such as wind, solar, battery, diesel, etc.

- IEC 61850-90-5: Synchrophasor transmission according to IEC 61850.

- IEC 61850-90-8: Modeling of Electric Vehicle Charging as per IEC 61850.

However, several ADN entities such as phasor measurement unit (PMU), controllable loads, distribution static compensator (DSTATCOM), electric vehicle (EV), solar home system (SHS) are not been modeled in the IEC 61850 standard. Therefore, several researchers have developed information models for all such entities of ADN which are not yet modeled. Authors in [4] have proposed information for IEC 61850-90-5 PMU and provided detailed comparison between the existing IEEE C37.118.2-based PMU and IEC 61850-90-5-based PMU. Performance evaluation in terms of latency, for different network scenario has been presented. Since the GOOSE and SV type messages are restricted to a local substation and hence cannot be transported in a wide area network (WAN) because of absence of IP layer, modified GOOSE and SV messages have been used. These are known as R-GOOSE (Routable GOOSE) and R-SV (Routable Sampled Values).

Due to absence of any logical node for controllable load, authors in [2] have proposed logical node CNLO by utilizing the generic logical node. The information model for controllable load is also presented. Modeling of Flexible AC transmission system devices is not yet proposed in the standard. Realizing this knowledge gap, authors in [18] have proposed DSTATCOM controller information model as per IEC 61850 standard. The specific information exchanges in the DSTATCOM controller are modeled as instance of logical nodes in their work.

The impact on EV on smart grid has been presented in literature [19–22]. Extending the charging support-based information model of EV presented in IEC 61850-90-8, authors in [23], have amended the current information model of EV in order to include the discharging functionality. The proposed information model in their work can support both G2V (Grid to Vehicle) as well as V2G (Vehicle to Grid) functionality.

Solar Home System (SHS) is a small energy system with a PV panel on its rooftop used for energy generation. The IEC 61850-based model of SHS and Smart Meters to manage tariff structure for bidirectional power transfer has been presented in [24].

4. Performance evaluation of ADN

Performance evaluation for a communication network is computed based on certain communication parameters which are discussed in this section.

4.1 Performance evaluation metrics

Various actors involved in ADN operation are geographically distantly located. In order to manage and control the ADN operation, a coordinated action among various ADN actors is necessary. This coordinated action can be realized by a foolproof communication for control and management of a ADN. To ensure this foolproof communication, there are certain communication parameters to which every ADN communication network must adhere. For effective operation, these parameters must be followed. They are as follows.

4.1.1 Network latency

Latency may be described as the delay on the transmitted data between various ADN components. Network latency is the time elapsed in transferring a data packet from source to destination in a communication network. The network latency is also known as End-to-End delay. In certain time critical ADN applications, network latency is not tolerable and a constraint on network latency is defined in communication standards. Applications such as wide-area situational awareness system, protection strategies are highly time critical and hence requires very low latency rates. For other ADN applications, such as data logging, etc., are not very time critical, network latency is tolerable and has larger acceptable limits.

4.1.2 Data delivery criticality

Data delivery criticality is defined based on the type of data which is communicated in a ADN communication network. Certain commands such as trip signals, critical alarms, etc., are critical in nature and requires guaranteed delivery to the destination. A ADN can operate in grid connected as well as islanded mode. To switch over from one ADN operation from one mode to another requires communication of commands from ADN central controller to point of common coupling circuit breaker. These commands are highly data critical commands and high data delivery criticality is to be ensured for safe operation of ADN.

4.1.3 Quality of service (QoS)

The communication between a power provider and the power consumer is a key aspect in ADN. Degradation in performance due to delay, network outage, jitter may affect the system reliability and thus a QoS mechanism is needed. A continuous cycle is required to achieve effective QoS in a system. QoS implementations in ADN communication networks are of paramount importance. Providing effective QoS in ADN communication network is becoming a prime aspect in today's enterprise of communication network.

A use case for QoS in a smart grid scenario can be considered is the steaming of various smart sensors for large scale Internet of Things project in smart buildings. These smart sensors collect data such as temperature, pressure, and humidity and are highly time critical. Thus, with effective QoS, this data can be efficiently identified, analyzed, marked and queued accordingly.

4.1.4 Interoperability

Interoperability may be defined as the ability of different information systems and technologies to seamlessly exchange data and interact with other system for required application. In order to realize capabilities of a ADN, technology deployments must connect large numbers of smart devices and systems involving hardware and software. These devices are manufactured from a wide range of vendors having little to much differences in their design and capabilities. Thus, these devices are vendor-specific and cannot be seamlessly integrated within one project. This hindrance creates inconvenience to the operator.

To effectively realize the ADN capabilities, interoperability is an important aspect of technology deployment and must be ensured. The prime requirement for a ADN communication network is on deployment of technologies having end to end integration with compatibility among them. A scheme that is driven and sustained by compliance with uniform standards is the prime motive in a ADN communication requirement.

4.1.5 Scalability

Scalability is defined as the capability of a system or network to handle increasing amount of work or sudden growth without any change on its performance. A use case of scalability can be a network switch with multiple devices connected on its ports. As the number of devices which are plugged-in increases, it is required that the performance of the network switch must remain the same. If there is no change in performance, it is said to be a scalable network switch otherwise it is non-scalable.

In a ADN communication network scenario, large number of DERs and other devices are connected to a ADN network. These devices rapidly add and delete in the network architecture. It is required that the performance of ADN communication network must not be altered with the changing network scenario.

4.1.6 Data security

An ADN communication network would be a wide area network and at times would use the resources of a public shared network such as Internet. Data security requires the transferred data to reach from the source to destination securely. The data must not be tampered in the communication path. An end to end cyber secure path must be achieved.

The cyberthreats are on rise in today's scenario and it is required to protect the power system automation data from cyberattacks. Data confidentiality, integrity must be maintained to provide effective services in ADN communication network. A use case can be a scenario in which a fault arises in a part of ADN and respective protection devices issues a trip signal to the breaker. This trip command in form of GOOSE message in an IEC 61850-based ADN must be communicated within a stipulated delay. Any intrusion would lead to tampering of data in such a manner that it becomes illegitimate and is of no use. Thus, failure of protection strategy would give rise to cascading fault which could lead to huge losses in terms of economy.

4.1.7 Reliability, robustness and availability

Reliability is defined as an attribute of a communication network that consistently performs according to the specifications. A communication network must not fail with increasing network traffic and performs consistently. Robustness is defined as the attribute of a communication network in which it is not vulnerable to any kind of faults and its performance is guaranteed. Availability of a communication network is defined as an attribute that a communication network device must be readily available at all times and there should be no denial of service (DoS).

In an ADN communication network, reliability, robustness and availability must be ensured. For ensuring functions such as protection, operation, management and control of ADN a highly reliable, deterministic, robust and available communication network must be developed.

4.1.8 Standardization

There is a great effort for standardizing the smart grid communication. The benefits of standardization include easy integration of devices, a holistic framework for working and application, a larger business perspective for new entrants. For cost effective and wide spread deployment of smart grid/ADN, interoperability and open interfaces for future extensions standardized solutions are a necessity. The standardization of ADN is driven by government's worldwide by defining new policies and framework. Also, a large number of standardization organizations from ICT and energy industry are considering ADN standardization a priority issue.

For ADN standardization there are various standardization committees working in this direction. Some of the major standardization players are International Electrotechnical Commission (IEC), National Institute of Standards and Technologies (NIST), ISO/IEC JTC1, German Commission for Electrical, Electronic and Information Technologies (DKE), etc. Due to presence of multiple vendor devices in a ADN communication network, the activities of standardization are necessary to provide seamless deployments. The existing smart grid standards are from multiple standardization organizations and have to be developed continuously to deal with changes within regulatory, technical, political and organizational aspects.

4.2 Performance evaluation tools

For carrying out performance evaluation of ADN communication architecture before its actual deployment in field, software tools are employed. These tools help to present the performance of the communication network in terms of latency, throughput, jitter, etc. An emulated system can be developed with the help of performance evaluation tools. The following software-based evaluation tools are used for testing communication configuration of ADN. OPNET/Riverbed Modeler [25], OMNET++ [26], OMNEST, NS2/3 [27] and Qualnet [28].

OPNET/Riverbed Modeler: It provides a comprehensive simulation environment for modeling the communication network and distribution network. It can be used to analyze the performance and behavior of communication network. It provides a Graphical User Interface (GUI) and uses C programming language. It is now known as Riverbed Modeler.

OMNET++: OMNET++ is a component based, modular and open architecture discrete event simulator framework used for simulation of communication networks. Eclipse-based simulation library is used for OMNET++ simulation. Latest version of OMNET++ is OMNEST.

Network simulator (NS2/3): It is discrete event simulator which provides substantial support for TCP simulations, routing simulations and multicast protocol simulation for wired and wireless network. It is an open source freely available software which is designed specifically for computer network simulation. NS3 is the advanced version of NS2.

Qualnet: It is a commercial (licensed) network simulator from scalable network technologies (SNT) which was initially developed for defense project. It is used to predict performance of wireless and wired networks and is a ultra-high fidelity software.

5. Cybersecurity in IEC 61850-based ADN

The IEC 61850-based communication for ADN relies on data transfer in terms of IEC 61850-based messages over a wide area network to realize various features of smart communication. This wide area network is usually a public network like Internet and hence demands proper message security. Ensuring cybersecurity in smart grids employing IEC 61850 requires implementation of IEC 62351's guidelines for different power system operations. However, IEC 62351 cannot handle communications in WANs with several nodes exchanging information at the same time. Therefore, scalability is the bottleneck of cybersecurity in smart grids. Mapping XMPP to IEC 61850 messages can solve this problem. Very recent IEC 61850-8-2 provides these mappings only for MMS messages. There is an immediate need to study feasibility and performance of these mappings. Furthermore, to fully implement XMPP in IEC 61850-based networks, it should be mapped to other IEC 61850 messages, namely R-GOOSE and R-SV. A study for the assessment of delays caused due to processing probabilistic signature scheme (PSS) as per IEC 62351-6 standard in IEC 61850-based GOOSE messages has been presented in [29]. It was concluded that the existing security scheme does not meet the time criticality of GOOSE messages and there is a need of amendment of IEC 62351-6 for providing effective cybersecurity and timing requirements for IEC 61850-based messages.

6. Conclusion

This chapter presents communication technologies and standards which can be employed in an ADN. Role of information and communication technology is inevitable in management of ADN. Thus, this chapter presents communication technologies and protocols which can be adopted for communication configuration of ADN. Among the existing communication protocols for ADN, IEC 61850 is found to be most suitable and acceptable worldwide for communication standardization of ADN. Also, software tools which are employed for simulation of communication architecture of ADN have been discussed. Also, cybersecurity needs for IEC 61850-based ADN has been discussed in this chapter.

Conflict of interest

Authors declare no conflict of interest.

Author details

Mohd Asim Aftab[1]*, S.M. Suhail Hussain[2] and Ikbal Ali[1]

1 Department of Electrical Engineering, Jamia Millia Islamia (A Central University), New Delhi, India

2 Fukushima Renewable Energy Institute, AIST (FREA), Koriyama, Japan

*Address all correspondence to: mohdasimaftab4@gmail.com

References

[1] Koutsoukis NC, Siagkas DO, Georgilakis PS, Hatziargyriou ND. Online reconfiguration of active distribution networks for maximum integration of distributed generation. IEEE Transactions on Automation Science and Engineering. 2017;**14**(2):437-448

[2] Ali I, Hussain SMS. Communication design for energy management automation in microgrid. IEEE Transations on Smart Grid. 2018;**9**(3):2055-2064

[3] Communication networks and systems for power utility automation-IEC 61850. 2nd ed. International Electrotechnical Commission; 2013

[4] Ali I, Aftab MA, Hussain SMS. Performance comparison of IEC 61850-90-5 and IEEE C37.118.2 based wide area PMU communication networks. Journal of Modern Power Systems and Clean Energy. 2016;**4**(3):487-495

[5] OpenADR [Internet]. 2019. Available from: http://www.openadr.org/ [Accessed: March 14, 2019]

[6] IEEE Application Guide for IEEE Std 1547, IEEE Standard for Interconnection Distributed Resources with Electric Power Systems-IEEE 1547. Institute of Electrical and Electronics Engineers. 2008

[7] Guide for Smart Grid Interoperability of Energy Technology and Information Technology Operation with the Electric Power System (EPS), and End-Use Applications and Loads-IEEE 2030. Institute of Electrical and Electronics Engineers. 2011

[8] Telecontrol Equipment and Systems-Part 5: Transmission Protocol. IEC 60870-5-101. International Electrotechnical Commission. 2018

[9] Ali I, Hussain SMS, Aftab A. Communication modeling of phasor measurement unit based on IEC 61850-90-5. In: Annual IEEE India Conference (INDICON). New Delhi; 2015. pp. 1-6

[10] IEEE Standard for Synchrophasor Measurements for Power Systems-IEEE C37.118.1. Institute of Electrical and Electronics Engineers. 2011

[11] IEEE Standard for Synchrophasor Data Transfer for Power Systems-C37.118.2. Institute of Electrical and Electronics Engineers. 2011

[12] Distributed Network Protocol (DNP) [Internet]. 2019. Available from: https://www.dnp.org/ [Accessed: March 14, 2019]

[13] Aftab MA, Roostaee S, Hussain SMS, Ali I, Thomas MS, Mehfuz S. Performance evaluation of IEC 61850 GOOSE based inter-substation communication for accelerated distance protection scheme. IET Generation Transmission and Distribution. 2018;**12**(18):4089-4098

[14] Ali I, Hussain SMS, Tak A, Ustun TS. Communication modeling for differential protection in IEC-61850-based substations. IEEE Transactions on Industry Applications. 2018;**54**(1):135-142

[15] Ustun TS, Ozansoy C, Zayegh A. Modeling of a centralized microgrid protection system and distributed energy resources according to IEC 61850-7-420. IEEE Transactions on Power Systems. 2012;**27**(3):1560-1567. DOI: 10.1109/TPWRS.2012.2185072

[16] Ustun TS, Ozansoy C, Ustun A. Fault current coefficient and time delay assignment for microgrid protection system with central protection unit. IEEE Transactions on Power Systems.

2013;**28**(2):598-606. DOI: 10.1109/TPWRS.2012.2214489

[17] Communication Networks and Systems for Power Utility Automation—Part 9-2: Specific Communication Service Mapping (SCSM)—Sampled Values Over ISO/IEC 8802-3, IEC 61850-9-2, Ed. 2.0. 2011

[18] Hussain SMS, Aftab MA, Ali I. IEC 61850 modeling of DSTATCOM and XMPP communication for reactive power management in microgrids. IEEE Systems Journal. 2018;**12**(4):3215-3225. DOI: 10.1109/JSYST.2017.2769706

[19] Ustun TS, Zayegh A, Ozansoy C. Electric vehicle potential in Australia: Its impact on smartgrids. IEEE Industrial Electronics Magazine. 2013;**7**(4):15-25. DOI: 10.1109/MIE.2013.2273947

[20] Hussain SMS, Ustun TS, Nsonga P, Ali I. IEEE 1609 WAVE and IEC 61850 standard communication based integrated EV charging management in smart grids. IEEE Transactions on Vehicular Technology. 2018;**67**(8):7690-7697. DOI: 10.1109/TVT.2018.2838018

[21] Ustun TS, Ozansoy CR, Zayegh A. Implementing vehicle-to-grid (V2G) technology with IEC 61850-7-420. IEEE Transactions on Smart Grid. 2013;**4**(2):1180-1187. DOI: 10.1109/TSG.2012.2227515

[22] Ustun TS, Hussain SMS, Kikusato H. IEC 61850-based communication modeling of EV charge-discharge management for maximum PV generation. IEEE Access. 2019;**7**:4219-4423. DOI: 10.1109/ACCESS.2018.2888880

[23] Aftab MA, Hussain SMS, Ali I, Ustun TS. IEC 61850 and XMPP communication based energy management in microgrids considering electric vehicles. IEEE Access. 2018;**6**:35657-35668. DOI: 10.1109/ACCESS.2018.2848591

[24] Hussain SMS, Tak A, Ustun TS, Ali I. Communication modeling of solar home system and smart meter in smart grids. IEEE Access. 2018;**6**:16985-16996. DOI: 10.1109/ACCESS.2018.2800279

[25] OPNET [Internet]. 2019. Available from: https://www.riverbed.com/in/products/steelcentral/steelcentral-riverbed-modeler.html

[26] Omnet++ [Internet]. 2019. Available from: https://www.omnetpp.org/

[27] NS2 [Internet]. 2019. Available from: https://ns2tutor.weebly.com/

[28] Qualnet [Internet]. 2019. Available from: https://web.scalable-networks.com/qualnet-network-simulator-software

[29] Farooq SM, Hussain SMS, Ustun TS. Performance evaluation and analysis of IEC 62351-6 probabilistic signature scheme for securing GOOSE messages. IEEE Access. 2019;**7**:32343-32351. DOI: 10.1109/ACCESS.2019.2902571

Chapter 5

Density-Aware Smart Grid Node Allocation in Heterogeneous Radio Access Technology Environments

Vahid Kouhdaragh, Daniele Tarchi
and Alessandro Vanelli-Coralli

Abstract

Smart grid (SG) is an intelligent enhancement of the conventional energy grid allowing a smarter management. In order to be implemented, SG needs to rely on a communication network connecting different node types, implementing the SG services, with different communication and energy requirements. Heterogeneous network (Het-Net) solutions are very attractive, gaining from the allocation of different radio access technologies (RATs) to the different SG node types; however, due to the heterogeneity of the system, an efficient radio resource optimization and energy management are a complex task. Through the exploitation of the most significant key performance indicators (KPIs) of the SG node types and the key features of the RATs, a joint communication and energy cost function are here defined. Through this approach it is possible to optimally assign the nodes to the RATs while respecting their requirements. In particular, we show the effect of different nodes' density scenarios on the proposed allocation algorithm.

Keywords: smart grid, wireless communications, heterogeneous networks, heuristic optimization

1. Introduction

Smart grid (SG) systems are characterized by the presence of several applications aiming at efficiently managing the energy grid. In order to do this, a smart grid communication network (SGCN) is implemented strictly coupled with the energy grid, which is able to interconnect the different nodes managing the energy grid applications. A typical SGCN scenario is characterized by the presence of different radio access technologies (RATs), with different communication configurations and characteristics able to support the SG communication requirements. However, wireless communications are now deployed for supporting different applications; hence an efficient resource allocation to support different types of SG nodes should be performed in order to maximize the resource efficiency while respecting to the different SG node type communication requirements, with a particular attention to data rate, delay, reliability, and security. For associating the nodes to the considered RATs, we propose to measure the suitability of the assignment toward a certain RAT of a given node type based on its communication requirements and RAT communication characteristics. To this aim, an appropriate communication cost function (CCF) is defined based on some KPIs as a function of node types and densities and RAT characteristics. At the same time, low-power communication is the key for the realization of reduced form factor SG nodes; to this aim a suitable energy cost function (ECF) is also defined for minimizing energy per transferred data bit. A novel approach to assign the node to different RATs based on jointly exploiting the CCF and the ECF is here proposed resulting in a heterogeneous network that is efficient in both energy and communication aspects. Cost function (CF), based on node communication requirements, node type densities, and RAT features, defines the percentage of each node type which should be allocated to each RAT. The numerical results show the advantage of the proposed node allocation approach to different RATs with respect to the separate CFs. Moreover, a variable number of nodes are considered for understanding the impact of the nodes' density in terms of allocation efficiency.

2. Literature review

In the literature there are several papers dealing with SGCN, the node type communication requirements, and RAT selection. The summary of some of the most important papers is given in this section. In [1] a complete research on advanced metering infrastructure (AMI) exploring how to link consumer data gathered by utilities and managing insufficient communication network resources is considered. As an outcome of [1], it is clear that several data relay nodes and aggregators are needed to collect data produced by smart meters (SMs). Moreover, the SM message gathering problem is considered, and a method to collect multiple SM information incoming at the data collector nodes in order to reduce protocol overhead is considered. In [2] the capacity of a backhaul network to support the distribution grid in SG is considered. Several communication technologies are taken into consideration for coping with the SG communication requirements for the backhaul, connecting customer data collection points to the CS. A multi-hop wireless communication architecture is proposed, and its capability in meeting the requirements of the backhaul link is assessed by simulations. Despite introducing several RATs that have been suggested to fulfill the communication requirements at the distribution level, it is still lacking a method to assign SMs to the RATs. A method showing the suitability of a given RAT with respect to the other can be useful in assigning the nodes to the different RATs. In [3] the distribution network implemented through WiMAX is considered, by taking into account the communication characteristics of different SG. An analysis of the communication requirements of SG specifically to the power grid distribution domain and the consumer domain is also performed. In [3], the authors measure the smart metering aggregator data rate and the quality of service (QoS) performance; WiMAX is used as the backhaul from aggregators to the control station. In [4], the exploitation of wireless communications for SG applications has been discussed; however the node type communication requirements have not been considered. Moreover, the resource allocation efficiency has not been considered in this work, and the scalability and performance analysis on LTE networks have been left as future work.

There are other general surveys on the communication architecture in SG. In [5] the network implementation challenges in the power system settings have been deeply studied. Another survey on the communication architecture in SG is [6] focusing on communication network requirements for the main SG applications in home area network (HAN), neighborhood area network (NAN), and wide-area network (WAN). For different communication standards and SG use cases, in [6], the authors propose to collect the information about different communication requirements for diverse SG applications, in the three different fields, i.e., HAN, NAN, and WAN. Hence, a method to support the usage of different SG implementations is considered. The US Department of Energy discusses the main issues in SG by presenting the most significant goals of the different node types [7]. Furthermore, the communication requirements of different SG node types (i.e., data rate, delay sensitivity, reliability, and security) are explained [8]. Although there is no unique solution for elaborating a certain RAT for SG, the SG node communication requirements give a high-level vision to SG communication network designer in order to design the optimized RAT [9].

With the aim of designing a reliable and secure heterogeneous network, load balancing methods have been introduced [10]. Round-robin method, which distributes the traffic evenly among all the available base stations, regardless of existing load and performance, is proposed in [11]. As it is obvious, this type of balancing, regardless of RAT characteristics and SG node communication requirements and their adoptability, results in an inefficient heterogeneous network [12]. Load balancing is implemented in a way that the new user load is assigned to the base station with the lower traffic. Another network balancing method is named predictive node method where all the available base stations are observed over time and the trends are analyzed. The load balance works by assigning the traffic to the base stations with the best performance in terms of energy and spectral efficiency. Managing such type of balancing is very complex in both hardware and software aspects. Moreover, this type of observation needs a cognitive process and sensing and finding that results in having higher delay [13].

Several studies have been also performed related to different communication network infrastructures and their performance. In [14] the authors focused on LTE uplink transmission scheme. Single-carrier frequency-division multiple access (SC-FDMA) is the multiple access technique adopted in the LTE uplink transmission scheme. Compared with orthogonal frequency-division multiple access (OFDMA), used in the LTE downlink transmission and WiMAX, SC-FDMA has a better performance in terms of peak-to-average power ratio and frame error rate due to its coherent "single-carrier" property and built-in frequency diversity. In [14], an overview of LTE and LTE uplink transmission is done. The technology behind the uplink transmission (i.e., SC-FDMA) is analyzed in depth. In [15, 16] the authors studied a SG test-bed based on GSM capable of load management [15], fault detection, and self-healing [16]. This test-bed allows the implementation of various protocols and methodologies, which can be used for investigating the problems in SG.

Assessing the different communication network reliabilities is an important issue which has not been studied a lot. Wireless sensor networks for smart grid applications using a case study on link reliability and node lifetime evaluations in power distribution systems are described in [17]. The authors introduce the main scenarios and design challenges of wireless sensor networks for SG applications. SG node reliability in wireless sensor networks for SG applications is assessed through specific studies based on field tests in power system. Moreover, the authors in [17] discuss the challenges due to the RATs and SG channel conditions. One of the most used approaches for defining the reliability of a RAT for different node types is described in [18, 19]. In this technique, by means of the most important RAT reliability criteria, such as buffer size, link usefulness, latency, node generating rate, system status changing, and packet loss probability, the reliability of different network types for a certain node type is done.

3. System model

In order to support the communication requirements of the SG node types by using different RATs, we aim at defining a method where the percentage allocation of each SG node type to the considered RATs is optimized. By a suitable cost function (CF), defined in terms of the communication requirements of all the SG node types and the RAT communication characteristics, it is possible to evaluate the suitability of RATs for the selected SG node types [20].

In this model it is assumed that the SG node data are buffered in aggregators, each one considering a specific type of node; the aggregators are then connected with a control station. The data generated by the nodes are gathered to the aggregators and the collectors using different short- and medium-range RATs. In the proposed model, we aim at maximizing the allocation efficiency to the different RATs. The radio resource allocation is done based on the node densities and RAT communication characteristics and features. The allocation is performed by using a suitable CF, defining a matching score between RAT features and node communication characteristics. Moreover, the node traffic changes as a function of node density. This is also considered giving interesting insights in the numerical results related to the effect of the densities of the different node types.

The proposed CF is based on a joint approach between a CCF, defined in terms of the communication requirements and characteristics, and an ECF defined in terms of energy consumption per bit. By noticing that the energy efficiency expressed in energy spent per bit is trading off with respect to the spectral efficiency, by using the CCF in combination of ECF, an efficient method is introduced to make a heterogeneous network for heterogeneous SG nodes in an efficient way in which a trade-off between communication requirements and energy savings is considered. In order to have a scalar output for mapping different types of RATs to support different SG node type communication requirements a desirability value is defined. The node allocation is performed with the aim of maximizing the desirability of the RATs with respect to the node characteristics. Smaller values of the CF stand for a higher importance of a given RAT for a given node type. In the proposed approach, the RATs not qualified to achieve the communication requirements of the nodes are omitted.

A SG environment is characterized by several types of nodes, having different characteristics and requirements. In this section a brief description of the main SG node types and their characteristics is given.

Advanced metering infrastructure (AMI) is considered as the backbone of the SG. It is composed by smart meter (SM) nodes that cooperate on the power demand controlling with the goal of optimizing the energy consumption. Besides, AMI utilizes the power distribution management indirectly as it reports the consumption to the control center in order to optimize the power consumption. Meter reading allows a utility to collect data from electric, gas, and water meters and transfer data to a CS for billing and analysis.

Plug-in hybrid electric vehicles (PHEVs) are a node type used in electric transportation applications for managing both electricity flows from vehicles to the power grid (vehicle to grid (V2G)) and from the power grid to the vehicles (grid to vehicle (G2V)). Electric transportation applications allow to receive information of vehicle battery state of charge and inform vehicles about electricity prices [7, 21].

Wide-area situational awareness (WASA) is used for implementing the awareness of the SG, in order to get information and react with respect to unwanted and unbalanced situations that may cause some problems to the electrical grid [7, 8, 22, 23].

Distributed grid management (DGM) allows to implement a smart management of the power distribution network. Within this context cyber-attacks or risky weather conditions are considered. These bi-directional communications are vital to accomplish power distribution. The real-time procedure of grid structure, automation control, and information communication and information management to monitor and control the distribution grid is possible by using DGM [5–7, 17].

Distributed energy resources (DERs) are used for managing the distributed electrical sources with an impact on the user generation plants and distributed energy storage sites [7, 22, 24].

The communication requirements of the considered node types are reported in **Table 1**.

(a) Smart meter parameters					
Smart meters	**Reporting time period**	**Packet size**			
	Every 15 min	125 bytes (i.e., 1000 bits)			
(b) SG node characteristics [6, 25]					
SG node type	**Average data size (bytes)**	**Reporting time period [S]**	**Latency [S]**	**Reliability (%)**	***Security***
AMI					
SMs infrastructure	125	900 min	15		High
Wide-area protection					
Adaptive islanding	4–157	0.1	<0.1	>99.9	High
Predictive under-frequency load shedding	4–157	0.1	<0.1	>99.9	
Wide-area control					
Wide-area voltage stability control	4–157	0.5–5	<5	>99.9	High
Facts and HVDC control	4–157	30–120	<120	>99.9	
Cascading failure control	4–157	0.5–300	<5	>99.9	
Pre-calculation transient stability control	4–157	30–120	<120	>99.9	
Closed-loop transient stability control	4–157	0.02–6	<0.1	>99.9	
Wide-area power oscillation damping control	4–157	0.1	<0.1	>99.9	
Wide-area monitoring					
Local power oscillation monitoring	>52	0.1	<30	>99.9	High
Wide-area power oscillation monitoring	>52	0.1	<0.1	>99.9	
Local voltage stability monitoring	>52	0.5–5	<30	>99.9	
Wide-area voltage stability monitoring	>52	0.5–5	<5	>99.9	
PMU-based state estimation	>52	0.1	<0.1	>99.9	

(b) SG node characteristics [6, 25]					
SG node type	**Average data size (bytes)**	**Reporting time period [S]**	**Latency [S]**	**Reliability (%)**	***Security***
Dynamic state estimation	>52	0.02–0.1	<0.1	>99.9	
PMU-assisted state estimation	>52	30–120	<120	>99.9	
PHEV					
Electric transportation (utility interrogates PHEV charge status)	>100	2–4 per PHEV per day (7 am–10 pm)	<15 S	>98	Relatively high
DERs					
Distribution customer	>25	2–6 per dispatch period per day	<5 S	>99.5	High
Storage (charge/discharge command from DAC to the storage)		(discharge: 5 am–9 am or 3 pm–7 pm; charge: 10 pm–5 am)			

Table 1.
SG node communication requirements.

As different RATs are supposed to be used in a Het-Net, in this study, LTE, GSM, and three different satellite-based communication systems have been considered, where the three main constellation types have been used (i.e., low Earth orbit, medium Earth orbit, and geostationary Earth orbit), while the reference communication system has been considered to be the DVB-S2/DVB-RCS2 for the downlink/uplink.

The characteristics of the considered RATs are given in **Table** 7 in which different parameters of each RAT for different scenarios are given.

4. Cost function-based allocation

In order to support the smart grid communication requirements of the different node types by using different RATs, it is needed to define a method able to assign the nodes of each SG node type to the RATs in an efficient way. We propose to use a suitably defined cost function of all SG node communication requirements and RAT communication characteristics. The cost function is modeled in a way that the SG node requirements and the RAT characteristics are matched for maximizing their suitability. The cost function minimization allows to find the optimal percentage of nodes for ach SG node type to be allocated to each RAT.

4.1 Cost function definition

The CF is composed of two jointly coupled cost functions: the communication cost function and the energy cost function. The CCF is characterized by some parameters defined as KPIs; among others we focused our attention on data rate, delay, reliability, and security [20]. The ECF is based on the energy consumption per bit and based on the consideration that there is a trade-off between the spectral efficiency and energy per bit at the transmitter side [26, 27]. By using a joint CCF and ECF, an efficient method is introduced to design a heterogeneous network for different SG nodes.

The aim of the CCF is to map the matching degree of the SG nodes' requirements and the RAT characteristics by setting a RAT desirability value for each SG node type. The CCF is defined between the ith node type and the jth RAT which can be defined as

$$CCF_{ij} = \frac{\sum_{q=1}^{N_{KPI}} \left(W_{q_i} \cdot N_{q_{ij}}\right)}{\sum_{q=1}^{N_{KPI}} W_{q_i}} \tag{1}$$

where N_{KPI} is the number of KPIs we are considering and W_{q_i} and $N_{q_{ij}}$ are the weight of the qth KPI for the node type i and the normalized value of the qth KPI when considering the jth RAT type and the ith node type, respectively [20]. Eq. (1) can be rewritten by considering the four KPIs previously introduced as [20]

$$CCF_{ij} = \frac{W_{R_i} \cdot N_{R_{ij}} + W_{D_i} \cdot N_{D_{ij}} + W_{REi} \cdot N_{REij} + W_{SEi} \cdot N_{SEij}}{W_{Ri} + W_{Di} + W_{REi} + W_{SEi}} \tag{2}$$

where $i = 1, \ldots,$ N and $j = 1, \ldots,$ F represent the node types and the RATs; W_{R_i} and $N_{R_{ij}}$ are the data rate weight and normalized value for user type i and RAT type j, respectively; W_{D_i} and $N_{D_{ij}}$ are the delay weight and normalized value for user type i and RAT type j, respectively; W_{REij} and N_{REij} are the weight and the normalized value for reliability; and W_{SEij} and N_{SEij} are the normalized values for reliability and security, respectively.

Through the definition of proper weight and normalized values for every KPI, it is possible to integrate in a simpler way. This approach is also convenient for those KPIs, such as the reliability and the security that cannot be defined directly in a quantitative way, while a class categorization is used.

4.1.1 KPIs

The data rate weight for the ith node type is defined as

$$W_{R_i} = \frac{R_i}{R^{max}} \tag{3}$$

where R_i is the data rate required by the ith node type and $R^{max} = \max\{R_1, R_2, \ldots, R_N\}$ is the maximum rate among all the SG node types. The data rate normalized value can be written as

$$N_{R_{ij}} = \frac{R_i}{R_j^{RAT}}$$

in which R_j^{RAT} is the jth RAT proportional data rate.

The latency corresponds to the end-to-end delay to send the generated data at the SG node to the CS including the processing time, the propagation delay, the payload time, and the buffering time. The delay weight for node i is defined as

$$W_{D_i} = 1 - \frac{D_i}{D^{max}} \tag{4}$$

where D_i is the delay requirement for node type i and $D^{max} = \max\{D_1, \ldots, D_n\}$ is the maximum value among D_i. The latency normalized value for node i when using the RAT j is defined as

$$N_{D_{ij}} = 1 - \frac{D_{ij}}{D_i}$$

where

$$D_{ij} = D_{ij}^B + \hat{\alpha}_{ij}\left(D_{ij}^{proc} + D_{ij}^{prop}\right) + D_{ij}^t$$

and [10]

$$\hat{\alpha}_{ij} = \min\{\alpha_{i1}, \ldots, \alpha_{ij}\}$$

D_{ij} is the overall delay including the buffering delay D_{ij}^B, the processing delay D_{ij}^{proc}, the propagation delay D_{ij}^{prop}, and the transmission delay D_{ij}^t, and

$$\alpha_{ij} = \frac{D_i}{D_{ij}^B + D_{ij}^{proc} + D_{ij}^{prop}}$$

is a coefficient that can be assigned by the designer to highlight the propagation and processing delay. This is an arbitrary option for the designer to highlight the latency effect. The propagation and processing time are multiplied by α_{ij} to reflect the RAT with high delay like satellite communication (i.e., GEO) that should have a lower CF value for the nodes with lower delay sensitivity necessities.

The reliability value in a RAT is not easy to be defined. There are lots of issues and parameters which should be evaluated in the different fields [19]. The reliability in a network is often defined in terms of network availability in an end-to-end connectivity. Reliability weight for each node type is defined as

$$W_{REi} = \frac{RE_i}{RE^{max}} \tag{5}$$

where RE_i is the target reliability value for node type i and $RE^{max} = \max\{RE_1, RE_2, \ldots, RE_N\}$. The required reliability of the different SG node types is categorized from high to fairly medium, and a numerical value is allocated to each one. These values are allocated as the weights to the SG reliability requirement values (**Table 2**).

The reliability normalized value for node i when using the RAT j can be defined by resorting to the mismatch probability (MMPR) concept which is introduced in [18]. In this method the ith node type packet generation period, λ_i, is considered as a variable of the system, and μ_j is the service rate of the jth RAT BS. The delay requirement of the node is considered as a sufficient value for the target MMPR and RAT latency and is D_{ij}. MMPR depends also on the packet loss probability, $P_{l_{ij}}$, that is usually considered equal to 0.01 in the literature [10, 28]. Hence, the MMPR between the SG node type i and the RAT j can be defined as.

$$MMPR_{ij} = 1 - \left(1 - P_{l_{ij}}\right)^2 \cdot \left(\frac{v_{ij}}{\lambda_i + v_{ij}}\right) \tag{6}$$

Reliability	High, 1–0.99999	Fairly high, 0.99999–0.9999	Medium, 0.9999–0.999	Fairly medium, 0.999–0.99
RE_i	0.8	0.6	0.4	0.2

Table 2.
Reliability values of KPI for the weight evaluation.

where [29]

$$\mathrm{v_{ij}} = \frac{1}{D_{ij}} \tag{7}$$

$$MMPR_{ij} = 1 - \left(1 - \frac{\left(\left(1-\frac{\lambda_i}{\mu_j}\right)\times\left(\frac{\lambda_i}{\mu_j}\right)^{K_j}\right)}{\left(1-\left(\frac{\lambda_i}{\mu_j}\right)^{K_j+1}\right)}\right)^2 \left(\frac{\frac{1}{D_{ij}^B+\alpha_{ij}\cdot\left(D_{ij}^{proc}+D_{ij}^{prop}\right)+D_{ij}^t}}{\lambda_i+\frac{1}{D_{ij}^B+\alpha_{ij}\cdot\left(D_{ij}^{proc}+D_{ij}^{prop}\right)+D_{ij}^t}}\right) \tag{8}$$

and K_j is he jth RAT BS buffer size. Hence.

$$\mathrm{N}_{RE\mathrm{ij}} = \frac{MMPR_{\mathrm{ij}}}{N^{max}} \tag{9}$$

where

$$N^{max} = \max\left\{\mathrm{MMPR_{i1}}, \mathrm{MMPR_{i2}}, \ldots, \mathrm{MMPR_{ij}}\right\} \tag{10}$$

for a certain node type i and different RATs.

Security in a network and assessing it is not a straightforward issue, and to the best of our knowledge, there is no work to evaluate it for a certain type of the RATs [19]. The SG node security requirements exist in the literature and are evaluated by these terms: high, very high, and medium. The reliability weight numeric values proportional to the quality scale of security are given as [10].

$$\mathrm{W_{SEi}} = \begin{cases} 1 \text{ if Node i security requirement is High} \\ 0.8 \text{ if Node i security requirement is Slightly High} \\ 0.6 \text{ if Node i security requirement is Medium} \\ 0.4 \text{ if Node i security requirement is Low} \\ 0.2 \text{ if Node i security requirement is Very Low} \end{cases}$$

To define the normalized value for each different type of nodes in the SG supported by different communication technologies, some RAT characteristics should be considered [10]. The main parts are the response time (RST), encryption policy (ENP), and RATs communication standards complexity (COMC). In **Table 3** the values of each security standard for a certain RATs are shown.

	RST	ENP	COMC
LTE	Very low	High	Very high
GSM	Low	Fairly high	Fairly high
SAT (LEO)	Fairly high	Fairly high	Very high
SAT (LEO)	High	Fairly high	High
SAT (LEO)	Very high	Fairly high	Fairly high

Table 3.
Fulfillment of each security criterion for the considered RATs [10].

Encryption is the procedure in which the data get twisted in a way that just the planned receiver could decrypt the message to get its information. Based on the ENP mode in a certain RAT and the complexity of the RAT, using symmetric and asymmetric cryptography model [10] [29], the value can be given to each parameter. The weights of ENP based on these algorithms are defined as w_{enj}, while $\mathrm{rep_j}$ indicates the number of consecutive encryption algorithms. The $\mathrm{ENP_j}$ is defined as.

$$\mathrm{ENP_j} = 5\frac{w_{enj}^{\left(1/rep_j\right)}}{w_{en}^{max}} \tag{11}$$

where the ENP value has been normalized to 5 (based on the defined value for security parameters); as it can be seen by increasing the number of consecutive encryption algorithms, $\mathrm{ENP_j}$ decreases significantly, indicating an increased security level in the system, where its default value is 1 (**Table 4**).

The security non-normalized value for RAT j can be defined as [10]

$$NSE_j = \frac{\alpha_{ENP} \cdot ENP_j + \alpha_{COM} \cdot COMC_j + \alpha_{RST} \cdot RST_j}{\sum_{SEC=1}^{3} \alpha_{SEC}} \tag{12}$$

where the parameter α_{SEC} represents the weight of the security KPIs, while α_{ENP}, α_{COM}, and α_{RST} are the encryption, complexity, and response time weights; by making a set, the normalized value for security is achieved from Eq. (12) and is shown in **Table 5** with $0 < \alpha_{SEC} < 1$.

To include energy part in our proposed method, energy per bit of information to noise power spectral density ratio, E_b/N_0, for RAT j is considered. The rationale is that using a specific communication configuration for RAT j causes having different $\mathrm{E_b/N_0}$ values. Let us recall the Shannon formula expressing the capacity of a given link:

$$R_b = B \cdot \log_2\left(1 + \frac{S}{N}\right) \tag{13}$$

where S and N are the signal and noise power, respectively. Thus, the link efficiency expressed in bits per Hz is

$$\eta_j = \frac{R_b}{B} = \log_2\left(1 + \frac{S}{N}\right), \tag{14}$$

and $S = E_b \cdot R_b$ and $N = N_0 \cdot B$ where N_0 is the noise spectral density. By using Eqs. (13) and (14), we have [5].

	RSA	DES	3DES	AES
Decryption velocity	Slowest	Slow	Very slow	Fast
Security weight: w_{enj}	4–5 least secure	3–4 not secure enough	2–3 adequate security	1–2 excellent security

Table 4.
Encryption algorithm weight mapping.

Node type	Data rate [bps]	Delay sensitivity [s]	Average packet generation period [s]	Security	Reliability
WASA1	5000	0.1	0.1	High	99.999–99.9999%
WASA2	8000	2.5	2.5	High	99.999–99.9999%
WASA3	3200	120	60	High	99.999–99.9999%
WASA4	5000	0.05	0.1	High	99.999–99.9999%
WASA5	1250	0.05	0.1	High	99.999–99.9999%
WASA6	1000	120	60	High	99.999–99.9999%
WASA7	2500	2.5	2.5	High	99.999–99.9999%
WASA8	15,000	15	15	High	99.999–99.9999%
WASA9	75,000	15	15	High	99.999–99.9999%
DGM1	10,000	0.1	1	High	99–99.999%
DGM2	5000	0.025	1	High	99–99.999%
DGM3	5000	0.1	1	High	99–99.999%
DGM4	250,000	0.15	1	High	99–99.999%
DERs	2400	3	4 × 3600	High	99–99.99%
PHEV	800	5	6 × 3600	Relatively high	99–99.99%

Table 5.
Node communication requirements.

$$\begin{aligned}\eta_j &= \log_2\left(1+\frac{E_b \cdot R_b}{N_0 \cdot B}\right)\\ &= \log_2\left(1+\frac{E_b \cdot \eta \cdot B}{N_0 \cdot B}\right)\\ &= \log_2\left(1+\frac{E_b \cdot \eta_j}{N_0}\right)\\ &= \log_2\left(1+\frac{E_b}{N_0}\eta_j\right)\end{aligned} \tag{15}$$

thus, we can state that

$$2^{\eta_j} = 1 + \frac{E_b}{N_0} \cdot \eta_j \tag{16}$$

If we rewrite $\frac{E_b}{N_0} = \eta en_j$, thus

$$\eta en_j = \frac{2^{\eta_j} - 1}{\eta_j} \tag{17}$$

For the same bandwidth, B, N_0 remains fixed since it depends on B and temperature K and Boltzmann coefficient, which are fixed. Hence, if the spectral efficiency changes, E_b or energy per information bit is changed. Thus, the different signal (in RATs) with different spectral efficiency can be compared in terms of energy efficiency. To do so, an algorithm should be applied. It can be defined in the following way:

$$Nb_j = \frac{\eta en_j}{\eta en^{max}} \tag{18}$$

where ηen_j is the set of $\frac{E_b}{N_0}$ for different RATs. Thus, $\eta en^{max} = \max\left\{\eta en_1, \dots, \eta en_j\right\}$ and $0 < Nb_j < 1$; hence the ECF when the node type i uses the RAT j can be defined as

$$\mathrm{ECF_{ij}} = \frac{\frac{2^{\eta_j}-1}{\eta_j}}{\eta \mathrm{en^{max}}}$$

4.2 Optimal allocation

At first it should be mentioned that allocation is based on the number of the nodes as a result of the node densities. The number of the nodes is achieved by multiplying the density of the node in the size of the area. For different scenarios and nodes, these values are given in **Table 6**. The certain type of the node traffic is achieved as a function of its density in a certain area. Using the previously defined CCF and ECF, it is possible to define an allocation rule β_{ij}^{C} is RAT j (communication) desirability value for node type i that shows the percentage of the node type i (using CCF) traffic which should be supported by RAT j and for a certain node type i, $\sum_{j=1}^{F}\beta_{ij}^{C} = 1$:

$$\beta_{ij}^{C} = \frac{(1 - CCF_{ij})}{\sum_{j=1}^{F}(1 - CCF_{ij})} \tag{19}$$

R,D node types	Scenarios									
	Scenario 1		Scenario 2		Scenario 3		Scenario 4		Scenario 5	
	R	*D*	*R*	*D*	*R*	*D*	*R*	*D*	*R*	*D*
1 AMI	3	500	5	50	4	300	2	3000	3	1000
2 PHEV	40	3000	30	2000	30	1000	30	3000	40	200
3 DERs	40	3000	40	2500	20	4000	40	2000	20	300
4 DGM1	1	4	1	5	2	1	1	4	1	2
5 DGM2	1	100	2	20	2	20	2	15	1	18
6 DGM3	1	100	2	15	2	25	6	6	3	5
7 DGM4	1	5	1	3	2	1	1	4	1	2
8 WASA1	3	10	2	12	3	2	5	1	4	2
9 WASA2	1	40	1	25	1	80	2	13	1	6
10 WASA3	1	2000	3	300	2	700	3	250	2	800
11 WASA4	1	8	2	2	1	10	2	3	2	3
12 WASA5	1	4	2	2	2	2	1	3	2	10
13 WASA6	3	200	1	300	2	250	2	150	4	100
14 WASA7	2	1	3	1	1	3	2	2	1	30
15 WASA8	5	10	4	8	3	7	3	9	4	60
16 WASA9	5	10	6	3	3	8	4	15	2	8

Table 6.
The different scenarios, for different node types, coverage area radius (R), and node densities (D).

Similarly β_{ij}^{E} is RAT j (energy aspect) desirability value that shows the percentage of the node type i (using ECF) traffic which should be supported by RAT j and for a certain node type i, $\sum_{j=1}^{F}\beta_{ij}^{E}=1$:

$$\beta_{ij}^{E}=\frac{\left(1-ECF_{ij}\right)}{\sum_{j=1}^{F}\left(1-ECF_{ij}\right)} \tag{20}$$

If we define with w_C and w_E the weights used for the node allocation based on CCF and ECF, respectively, showing the importance of communication and energy aspects of node (they can depend on designer goals; in this paper $w_\eta=w_\beta=1$), P_{ij} is the percentages of the node type i that is assigned to RAT j based on both CCF and ECF values and can be rewritten as

$$\begin{aligned}
P_{ij}&=\frac{w_C\beta_{ij}^{C}+w_E\beta_{ij}^{E}}{w_C+w_E}\\
&=\frac{w_C\dfrac{\left(1-CCF_{ij}\right)}{\sum_{j=1}^{F}\left(1-CCF_{ij}\right)}+w_E\dfrac{\left(1-ECF_{ij}\right)}{\sum_{j=1}^{F}\left(1-ECF_{ij}\right)}}{w_C+w_E}\\
&=\frac{w_C\cdot\dfrac{\left(1-\dfrac{\sum_{q=1}^{N_{KPI_u}}\left(W_{qi}\cdot N_{q_{ij}}\right)}{\left(\sum_{q=1}^{N_{KPI_u}}W_{q_i}\right)}\right)}{\sum_{j=1}^{F}\left(1-\dfrac{\sum_{q=1}^{N_{KPI_u}}\left(W_{qi}\cdot N_{q_{ij}}\right)}{\left(\sum_{q=1}^{N_{KPI_u}}W_{q_i}\right)}\right)}+w_E\cdot\dfrac{\left(1-\dfrac{\dfrac{2^{\eta e_j}-1}{\eta e_j}}{\eta en^{max}},\right)}{\sum_{j=1}^{F}\left(1-\dfrac{\dfrac{2^{\eta e_j}-1}{\eta e_j}}{\eta en^{max}},\right)}}{w_C+w_E}
\end{aligned} \tag{21}$$

It should be mentioned that by increasing the density of the nodes, the generated traffic is increased which affect the weights and normalized value of the CF in KPIs.

5. Numerical results

Considering an average number of nodes and collectors per branch defined by UTC, the numerical results based on the first proposed method show that regarding the number of KPIs which are used in the CF, selecting the best RATs for each type of SG nodes in a way that all the SG node communication requirements were fulfilled while the resource allocation done in an efficient, are changed.

Tables 6 and 7 showed the RAT, node density, and area size, respectively, for five different scenarios. In **Table 7** the RAT characteristics in terms of goodput, spectral efficiency (SE), coding rate and forward error correction (FEC), packet

		LTE	GSM	LEO	MEO	GEO
Scenario 1	Goodput	0.9	0.9	0.9	0.9	0.9
	SE	5.4	1.35	1.87	1.25	1.87
	Modulation	64 QAM	GMSK	4 PSK	4 PSK	8 PSK
	FEC rate	0.9	7/8	5/6	3/4	2/3
	PLP	0.001	0.007	0.009	0.01	0.015
	RTT	0.001	0.009	0.025	0.150	0.350
	Process	0.001	0.002	0.003	0.004	0.005
	Encryption	AES	DES	3 DES	AES	AES
Scenario 2	Goodput	0.9	0.9	0.9	0.9	0.9
	SE	4.8	1.35	2.9	3.2	3.9
	Modulation	32 QAM	GMSK	8 PSK	16 PSK	32 PSK
	FEC rate	1	0.95	8/9	7/8	5/6
	PLP	0.01	0.01	0.03	0.01	0.1
	RTT	0.005	0.009	0.025	0.150	0.350
	Process	0.001	0.002	0.003	0.003	0.003
	Encryption	AES	RSA	DES	3 DES	AES
Scenario 3	Goodput	0.9	0.9	0.9	0.9	0.9
	SE	2.9	2.4	2.07	3.1	1.87
	Modulation	8 QAM	8 PSK	8 PSK	16 PSK	4 PSK
	FEC rate	1	8/9	2/3	4/5	7/8
	PLP	0.001	0.02	0.02	0.04	0.05
	RTT	0.008	0.009	0.025	0.150	0.350
	Process	0.001	0.004	0.003	0.002	0.003
	Encryption	DES	DES	RSA	3 DES	AES
Scenario 4	Goodput	0.9	0.9	0.9	0.9	0.9
	SE	4	1.5	1.60	2	3
	Modulation	16 QAM	4 PSK	PSK	4 PSK	8 PSK
	FEC rate	1	8/9	7/8	3/4	2/3
	PLL	0.001	0.001	0.02	0.03	0.04
	RTT	0.005	0.008	0.025	0.150	0.350
	Process	0.005	0.005	0.005	0.005	0.005
	Encryption	AES	3 DES	AES	AES	AES
Scenario 5	Goodput	0.9	0.9	0.9	0.9	0.9
	SE	4	4	4	4	4
	Modulation	16 PSK	16 PSK	16 PSK	16 PSK	16 PSK
	FEC rate	1	8/9	3/4	3/4	3/4
	PLL	0.01	0.01	0.01	0.01	0.01
	RTT	0.005	0.005	0.025	0.150	0.350
	Process	0.001	0.001	0.001	0.001	0.001
	Encryption	AES	3 DES	DES	AES	3 DES

Table 7.
The different RAT features for the defined scenario.

loss probability (PLP), and round trip time (RTT) are given; complexity of the Radio Access Technologies, encryption at data link layer, RAT access point buffer size, access method, and other communication parameters are given, whose values are used to define CFs.

In Scenario 1, presented in **Table 7**, LTE and GSM have the maximum and the minimum SE, respectively. As depicted in **Figure 1**, due to the 64 QAM modulation scheme that allows to achieve the lowest CF and the reduced densities of the nodes, the LTE has the lowest CF, making it the preferable for Scenario 1. It is worth to be noticed that node types 5–8 and 11 and 12 cannot support the MEO and GEO satellite communications due to the strict latency requirements, while LEO is not supported only by 5 and 6. In general it can be noticed that LTE is the best choice for the CCF, while GSM and LEO are the second choice. In **Figure 2** the ECF is shown where it is possible to see that the highest cost is for the LTE, while the other three RATs have a similar behavior in terms of ECF; it affects the ECF based node allocation that foresees a lower amount of nodes allocated to the LTE. **Figure 3** shows instead the node assignment percentage to different RATs based on the CCF that reflect the CCF values depicted in **Figure 1**, by assigning more nodes to the RATs with a lower CCF value. Finally, in **Figure 4**, the joint CCF and ECF are

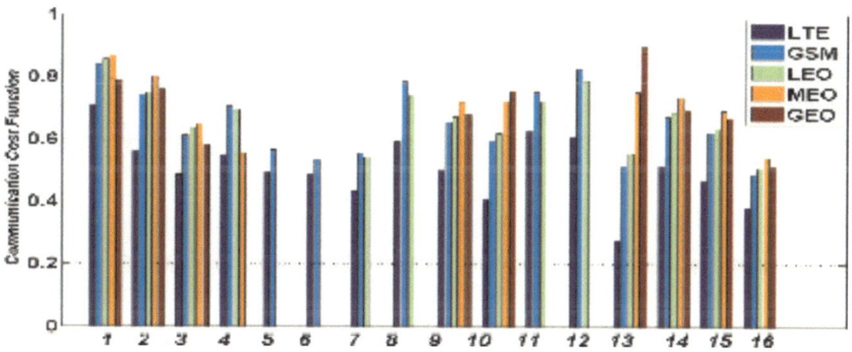

Figure 1.
CCF in S1.

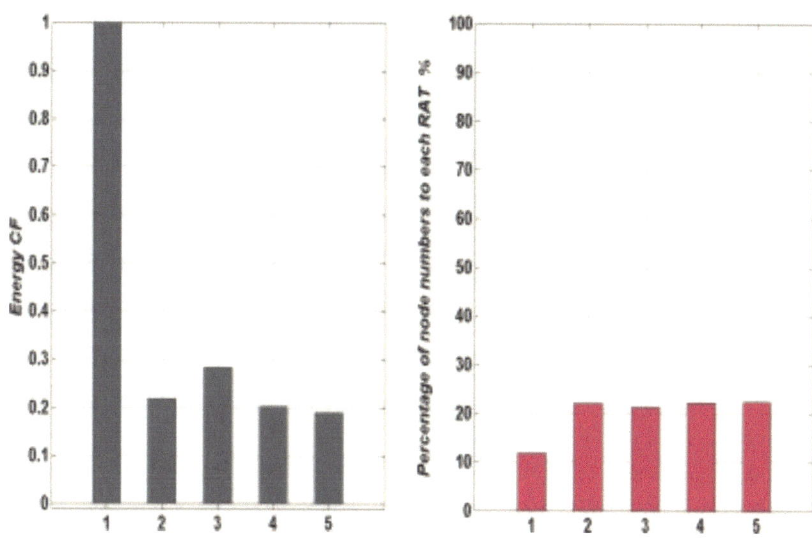

Figure 2.
ECF value and allocation percentage in S1.

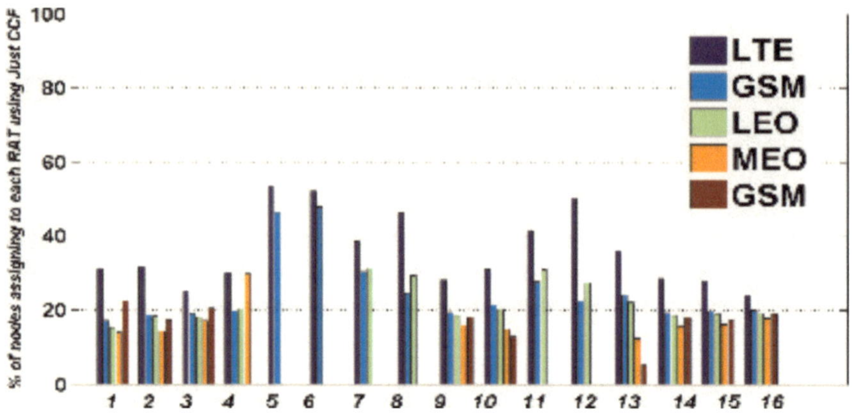

Figure 3.
Node allocation based on CCF in S1.

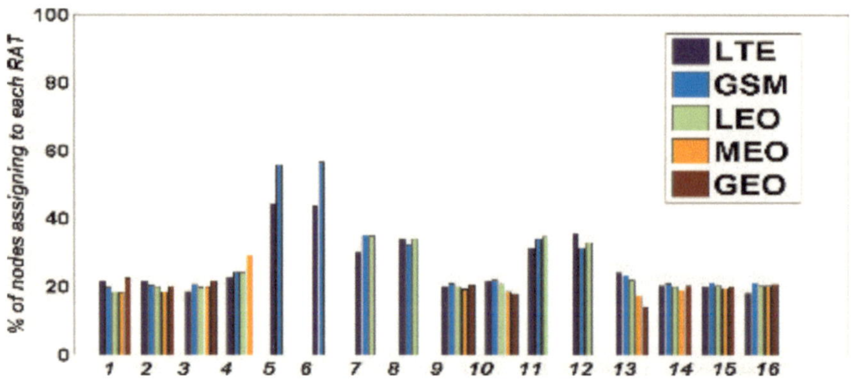

Figure 4.
Node allocation based on the joint CF in S1.

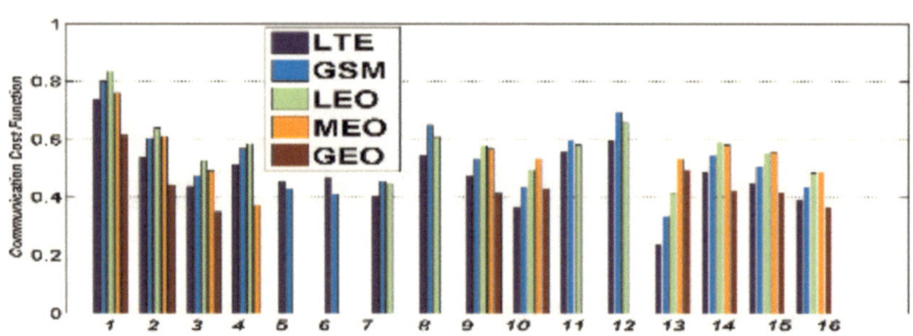

Figure 5.
CCF in S2.

considered for the node assignment; it is possible to notice both communication and energy effects on the nodes' assignment.

Scenario 2 represents a low-density node area. Also in this case, it is possible to notice that node types from 5 to 8 and 11 and 12 cannot use the satellite MEO and GEO, while LEO can be used for node types 7 and 8. By comparing SMs in Scenario 2 with SMs in Scenario 1, it is possible to notice that in Scenario 2 MEO has a better behavior than LEO and GSM, due to a higher SE. A similar behavior is noticed also for node types 2, 3, and 9. For the node type 4 in Scenario 2, we can instead notice a different behavior. The MEO is the best RAT in terms of CCF, but GSM has a lower CCF than the same node in Scenario 1. However, MEO with higher SE joined with its higher delay causes it to be a better choice than LTE. For nodes 5 and 6 in Scenario 2, though still LTE SE is higher than GSM SE, GSM with higher RTT and handling time makes GSM better in terms of CCF, while in nodes 7 and 8, CCF is slightly different if compared with the same node in Scenario 1, but, even with lower LTE SE, due to the density of the nodes, it has a better behavior in Scenario 2 than Scenario 1. For node 10, GEO CCF is lower than GSM, LEO, and MEO, but in node 13, GEO CCF is just lower than MEO. By looking at the node densities in these two scenarios, node 13 is lower. Higher density in node 10 rises the CFs for RATs with lesser SE. The RAT priorities for nodes 14, 15, and 16 are similar but are different compared with Scenario 1 because of different node densities using different communications in S2 (**Figure 5**). With the same ECF method, node assigning based on the CCF and based on both CCF and ECF is depicted in **Figure 6**.

In Scenario 3, it is worth to mention that in node type 7 (due to a high number of nodes whose data rate requirement is high) the SE and ECF weights are reduced (**Figures 7** and **8**). The other nodes have a similar behavior than the previous scenarios. The impact of the lower modulation order considered in Scenario 3 is clearly seen in the CCF differences and in node 16 CCFs (**Figures 9–12**).

In Scenarios 4 and 5, a similar behavior can be noticed. Though the RATs in these scenarios have different communication characteristics and, also, the node

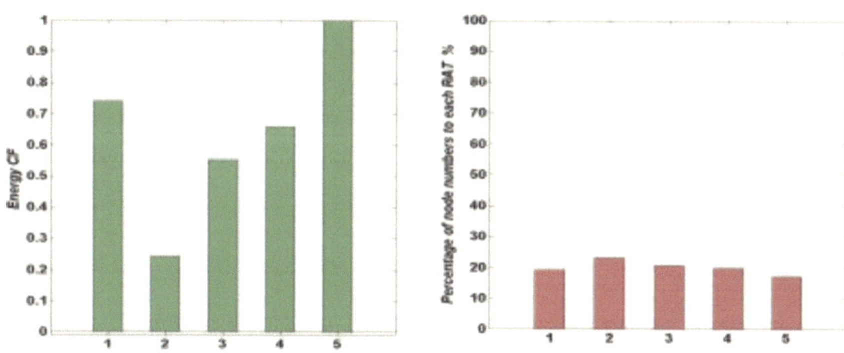

Figure 6.
ECF value and allocation percentage in S2.

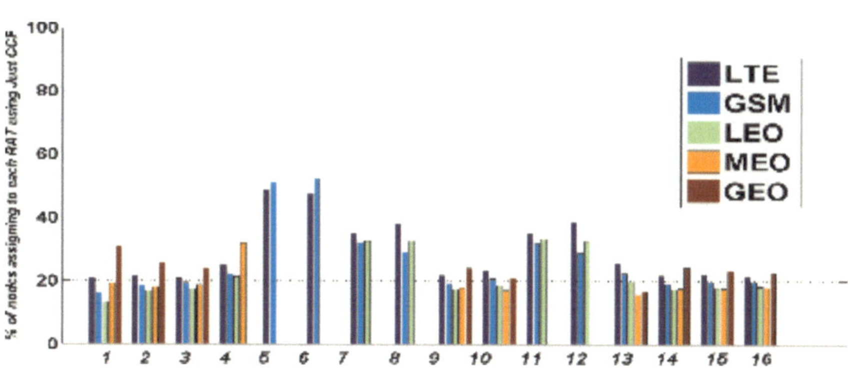

Figure 7.
Node allocation based on CCF in S2.

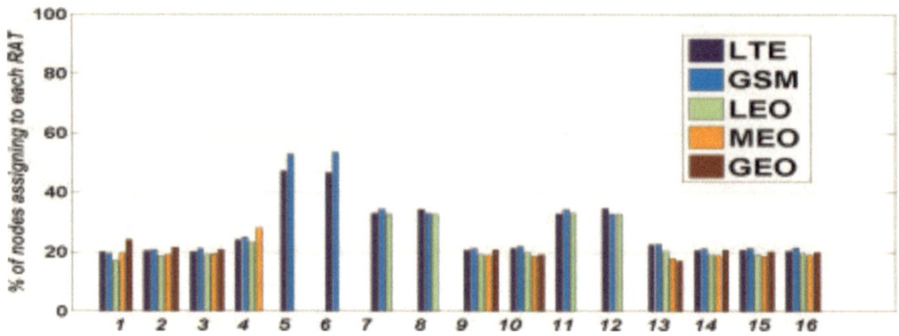

Figure 8.
Node allocation based on the joint CF in S2.

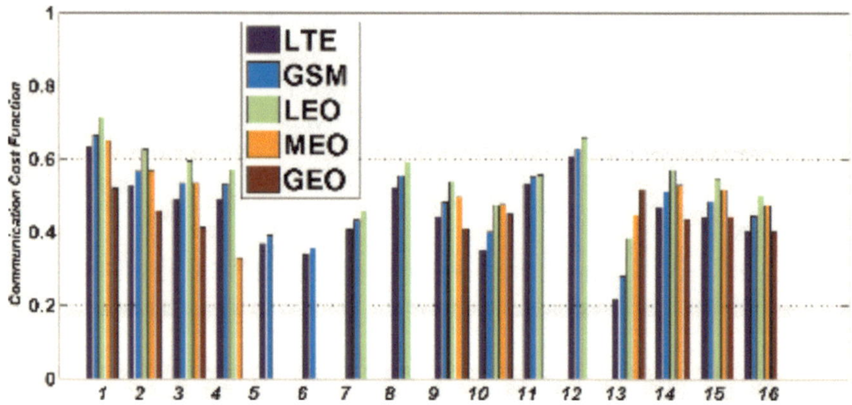

Figure 9.
CCF in S3.

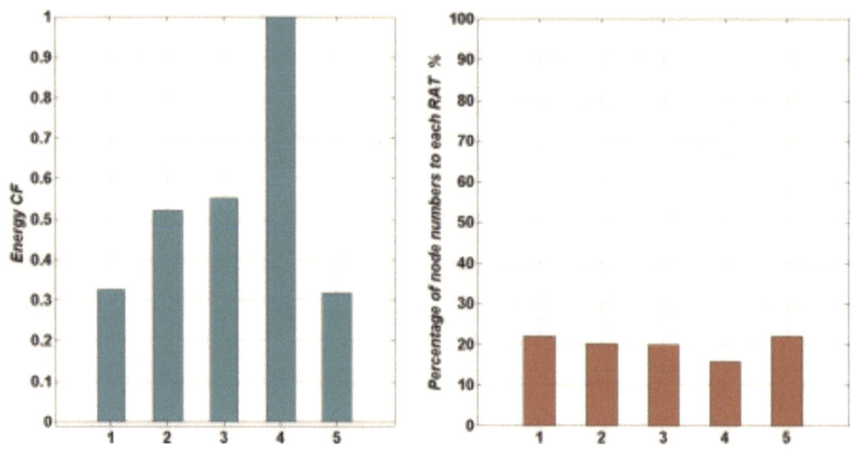

Figure 10.
ECF value and allocation percentage in S3.

densities and numbers are different, a similar behavior of RAT importance for different node types is present, although some important changes can be noticed. In Scenarios 4 and 5, node types 1, 2, 3, 4, 5, 6, 9, 11, 12, 13, 14, and 16 show the same behavior in terms of RATs when considering the CCF. For nodes 1–3, LTE is the

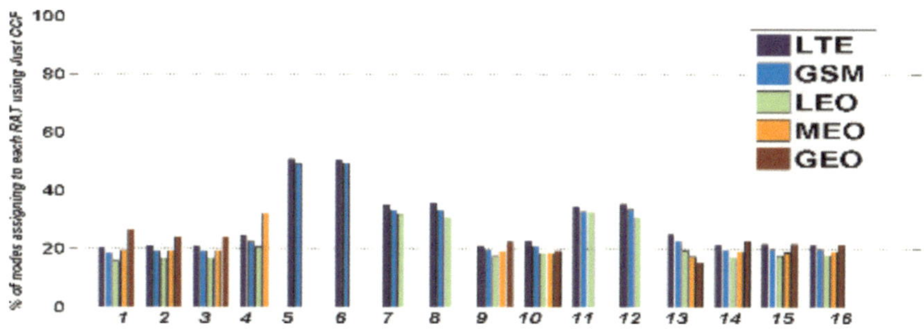

Figure 11.
Node allocation based on CCF in S3.

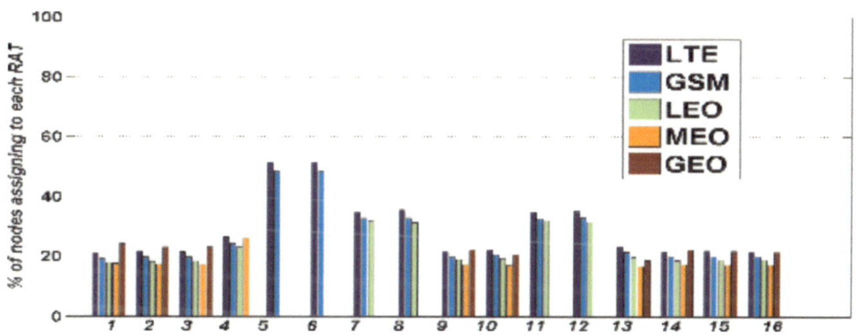

Figure 12.
Node allocation based on the joint CF in S3.

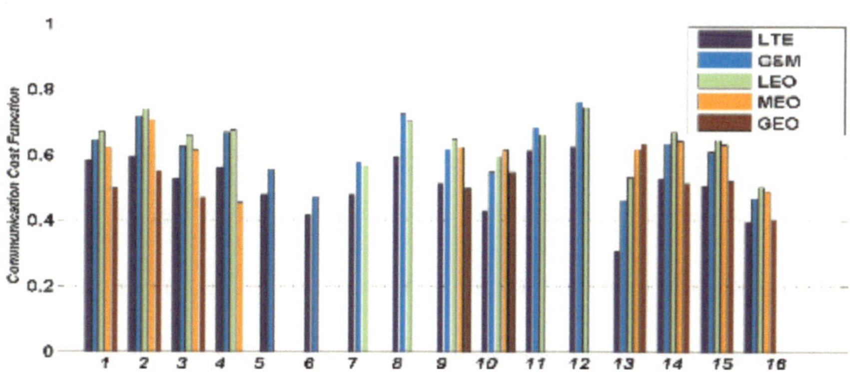

Figure 13.
CCF in S4.

best choice when considering the CCF, while in the case of node 4, MEO is the best selection. Node 4, due to its high data rate requirement and high delay sensitivity, cannot be supported by GEO.

In Scenario 4, the high PLP in GEO RAT type causes to have a lower reliability. Moreover, low PLP in MEO causes to have lower required data rate (i.e., node 13) resulting in having a lower CCF. Because of the high amount of data rate required by nodes 5 and 6 and due to the similar latency in GSM and LTE, LTE is better than GSM. In S4 node types 4, 5, 6, 9, 13, and 14, **Figures 13–16** show the same behavior in the sense of RATs used by the proposed CCF. The level of encryption, which is

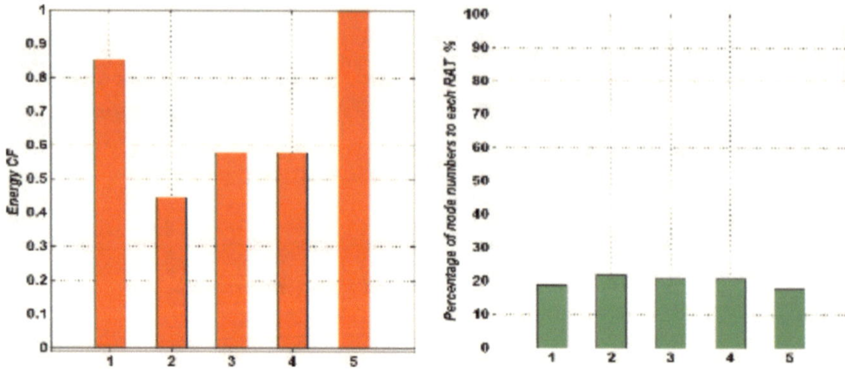

Figure 14.
ECF value and allocation percentage in S4.

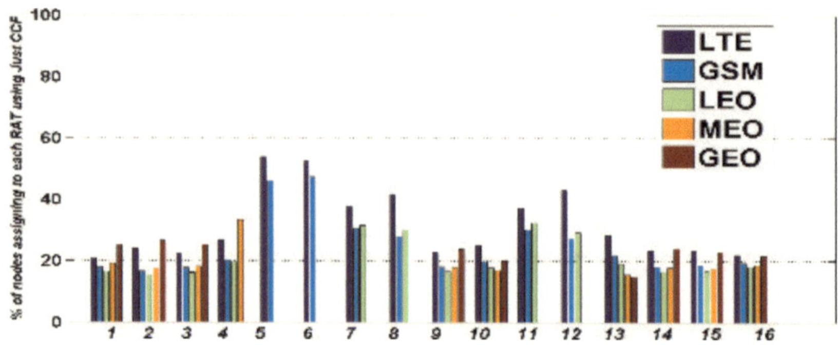

Figure 15.
Node allocation based on CCF in S4.

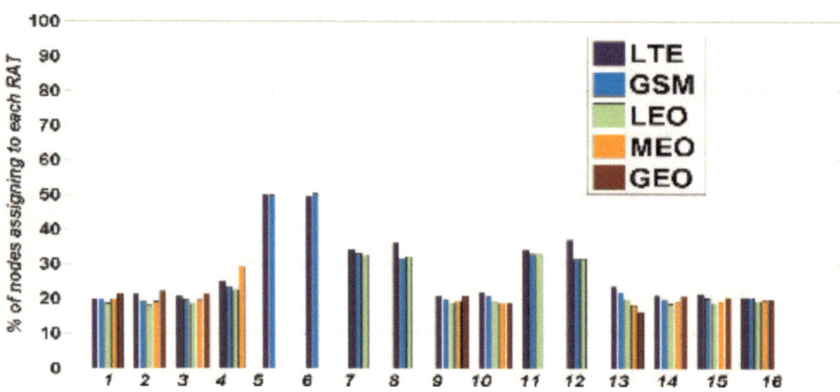

Figure 16.
Node allocation based on the joint CF in S4.

used as a part of security KPIs, is another criterion changing the node allocation strategy. In S5, for the node type 8, GSM has a lower CCF than LEO. Because of its encryption algorithm and coding rate which is better than LEO RAT and also delay fitting degree of LEO is improved, it has a lower CCF (**Figures 17–20**).

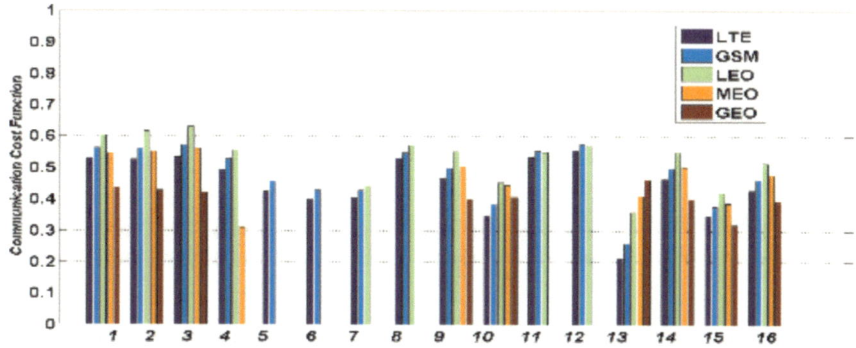

Figure 17.
CCF in S5.

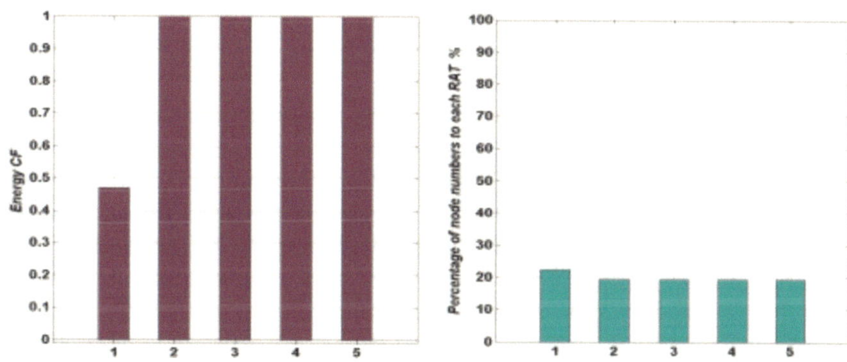

Figure 18.
ECF value and allocation percentage in S5.

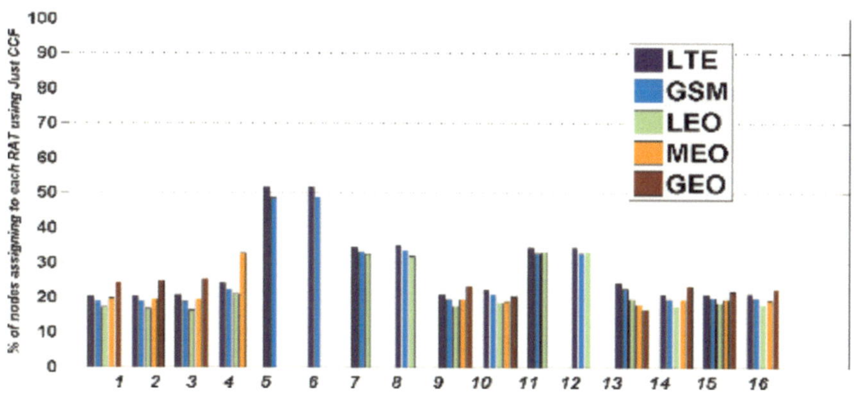

Figure 19.
Node allocation based on CCF in S5.

As it is shown, changing the RATs factor and its configuration consequences in changing the allocating percentage of the nodes to different RATs. For instance, in the different defined scenarios, the high-level modulation does not make the RAT as the priority. It means lots of well-defined criteria as the input of CCF define the popularity value of RATs for a certain node in a certain scenario.

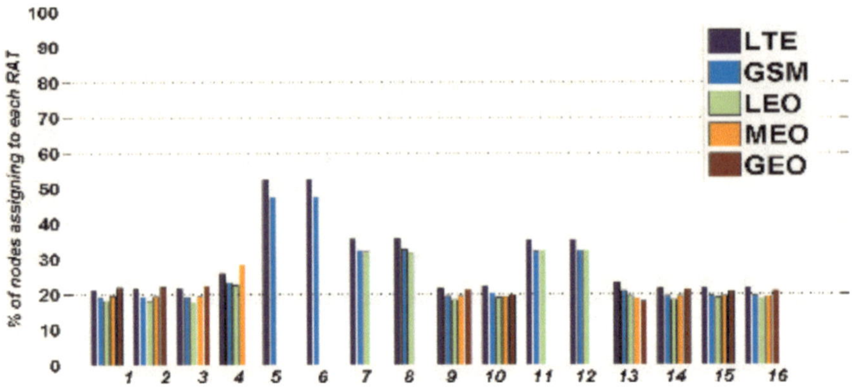

Figure 20.
Node allocation based on the joint CF in S5.

6. Conclusions

Different SG node types need to interact with a centralized management center by exchanging data with different requirements. Such communication can be implemented through different RATs, having different characteristics supporting the SG node communication requirements. A high-level fitness function is here defined aiming at matching the RAT communication characteristics and SG node type communication requirements. To this aim a joint communication and energy cost function is introduced for evaluating the effectiveness of the RATs when supporting different SG node types. Through the CF, the fraction of SG node types to be assigned to different RATs is obtained. The solution allows to achieve advantages in terms of load balancing and resource allocation. Thanks to the interesting results, this method can be extended to the inclusion of novel communication paradigms, such as those based on the presence of unmanned aerial vehicles (UAVs) as well as to the IoT-based communication paradigms.

Author details

Vahid Kouhdaragh, Daniele Tarchi* and Alessandro Vanelli-Coralli
Department of Electrical, Electronic and Information Engineering, University of Bologna, Bologna, Italy

*Address all correspondence to: daniele.tarchi@unibo.it

References

[1] Karimi B, Namboodiri V, Jadliwala M. On the scalable collection of metering data in smart grids through message concatenation. In: 2013 IEEE International Conference on Smart Grid Communications (SmartGridComm). 2013. pp. 318-323

[2] Karimi B, Namboodiri V. On the capacity of a wireless backhaul for the distribution level of the smart grid. IEEE Systems Journal. 2014;**8**(2):521-532

[3] Rengaraju P, Lung C-H, Srinivasan A. Communication requirements and analysis of distribution networks using WiMAX technology for smart grids. In: 2012 8th International Wireless Communications and Mobile Computing Conference (IWCMC). 2012. pp. 666-670

[4] Parikh PP, Kanabar MG, Sidhu TS. Opportunities and challenges of wireless communication technologies for smart grid applications. In: IEEE PES General Meeting. Providence, RI, USA; 2010

[5] Wang W, Xu Y, Khanna M. A survey on the communication architectures in smart grid. Computer Networks. Oct 2011;**55**(15):3604-3629

[6] Kuzlu M, Pipattanasomporn M, Rahman S. Communication network requirements for major smart grid applications in HAN, NAN and WAN. Computer Networks. 2014;**67**:74-88

[7] U.S. Department of Energy. Communications requirements of smart grid technologies. Report. 2010

[8] Deshpande JG, Kim E, Thottan M. Differentiated services QoS in smart grid communication networks. Bell Labs Technical Journal. 2011;**16**(3):61-81

[9] Gungor VC et al. A survey on smart grid potential applications and communication requirements. IEEE Transactions on Industrial Informatics. 2013;**9**(1):28-42

[10] Kouhdaragh V. A reliable and secure smart grid communication network using a comprehensive cost function. Journal of Energy and Power. 2017;**11**: 115-126

[11] Andrews JG, Singh S, Ye Q, Lin X, Dhillon HS. An overview of load balancing in het-nets: Old myths and open problems. IEEE Wireless Communications. 2014;**21**(2):18-25

[12] Yanmaz E, Tonguz OK. Dynamic load balancing and sharing performance of integrated wireless networks. IEEE Journal on Selected Areas in Communications. June 2004;**22**(5):862-872

[13] Kouhdaragh V. Optimization of smart grid communication network in a het-net environment using a cost function. Journal of Telecommunications. Nov 2016;**35**(1)

[14] Ghosh A, Zhang J, Andrews JG, Muhamed R. Fundamentals of LTE. Prentice Hall; 2011

[15] Nithin S, Radhika N, Vanitha V. Smart grid test bed based on GSM. Procedia Engineering. 2012;**30**:258-265

[16] Cypher DE, Golmie NT. NIST priority action plan 2, guidelines for assessing wireless standards for smart grid applications. In: NIST Interagency/ Internal Report (NISTIR) 7761. 2011

[17] Tuna G, Gungor VC, Gulez K. Wireless sensor networks for smart grid applications: A case study on link reliability and node lifetime evaluations in power distribution systems. International Journal of Distributed Sensor Networks. 2013;**2013**

[18] Shawky A, Olsen R, Pedersen J, Schwefel H. Class-based context quality

optimization for context management frameworks. In: 2012 21st International Conference on Computer Communications and Networks (ICCCN). Munich, Germany. 2012

[19] Tobgay S. Dynamic and Reliable Information Accessing and Management in Heterogeneous Wireless Networks. Center for TeleInFrastruktur (CTIF). Aalborg Universitet; 2013

[20] Kouhdaragh V, Tarchi D, Vanelli-Coralli A. A reliable, secure, and energy efficient smart grid node allocation algorithm for heterogeneous network scenarios. International Journal of Communication Systems. 2018

[21] Kemal M, Olsen R. Analysis of timing requirements for data aggregation and control in smart grids. In: 2014 22nd Telecommunications Forum Telfor (TELFOR). Belgrade; 2014. pp. 162-165

[22] Xiao Y. Communication and Networking in Smart Grids. CRC Press; 2012

[23] Hoag JC. Wide-area smart grid situational awareness communications and concerns. In: 2012 IEEE Energytech. Cleveland, OH; 2012

[24] Tseng CC, Wang L, Kuo CH. Application of hybrid mixing CDMA/IDMA/OCDMA/OIDMA for smart grid integration of renewable-energy resources. In: 2016 International Symposium on Computer, Consumer and Control (IS3C). Xi'an; 2016. pp. 878-882

[25] Communications: The Smart Grid's Enabling Technology. Cooperative Research Network Tech. Rep.; January 2015

[26] Lohier S, Rachedi A, Ghamri-Doudane Y. A cost function for QoS-aware routing in multi-tier wireless multimedia sensor networks. In: Pfeifer T, Bellavista P, editors. Wired-Wireless Multimedia Networks and Services Management. MMNS 2009. Lecture Notes in Computer Science. Vol. 5842. Berlin, Heidelberg: Springer; 2009

[27] Apsel A, Wang X, Dokania R. Approaches to low power radio design. In: Design of Ultra-low Power Impulse Radios. Analog Circuits and Signal Processing. Vol. 124. New York, NY: Springer; 2014

[28] Venkateswaran R, Sundaram DV. Information security: Text encryption and decryption with poly substitution method and combining the features of cryptography. International Journal of Computer Applications. 2010;**3**:28-31

[29] Kouh Daragh V. A Heterogeneous Communications Network for Smart Grid by Using the Cost Functions [PhD Dissertation Thesis]. Bologna, Italy: Alma Mater Studiorum Università di Bologna; 2017

Chapter 6

Environmental Impact of Information and Communication Equipment for Future Smart Grids

Vedad Mujan and Slavisa Aleksic

Abstract

The realization of the smart grid will require a deployment of additional information and communication technology (ICT) equipment in various domains but mostly the customer and distribution domains. All of these ICT equipment will unavoidably lead to an increase in electricity consumption and consequently to increased environmental sustainability issues and thus an overall environmental sustainability analysis if the future smart grid has to be performed. In order to obtain a meaningful environmental sustainability analysis, additionally to the operation phase, various other ICT equipment life cycle stages, i.e., raw material extraction and processing, manufacturing and assembly, recycling and disposal, as well as transportation, have to be included in the assessment as well. This chapter addresses the environmental sustainability of ICT equipment for smart grids involved in the advanced metering infrastructure (AMI) and home area network (HAN) smart grid applications. The environmental sustainability is analyzed by means of the exergy-based life cycle assessment (E-LCA) that is based on the second law of thermodynamics and takes the entire lifetime of ICT equipment into consideration. Some selected results of the E-LCA study are briefly presented and discussed. They have shown that the environmental impact of the additional ICT equipment cannot be neglected and has to be taken into account when assessing the environmental overall sustainability of smart grids.

Keywords: smart grids, advanced metering infrastructure (AMI), home area network (HAN), information and communication technology (ICT), exergy-based life cycle assessment (E-LCA), environmental sustainability

1. Introduction

The global energy demand has continuously been increasing over the last years and is expected to increase further at an average of 1.5% per year until 2040 [1]. The strongest increase is observed in countries which do not belong to the Organisation for Economic Co-operation and Development (OECD), known as non-OECD countries. The demand for energy in those countries is mainly caused by a strong economic growth, but also the growth in population has a remarkable contribution to this fact [1]. As opposed to this, most OECD countries have a slower economic growth, and also the growth in population in those countries is not that significant compared to non-OECD countries. **Figure 1** illustrates this development. Based on

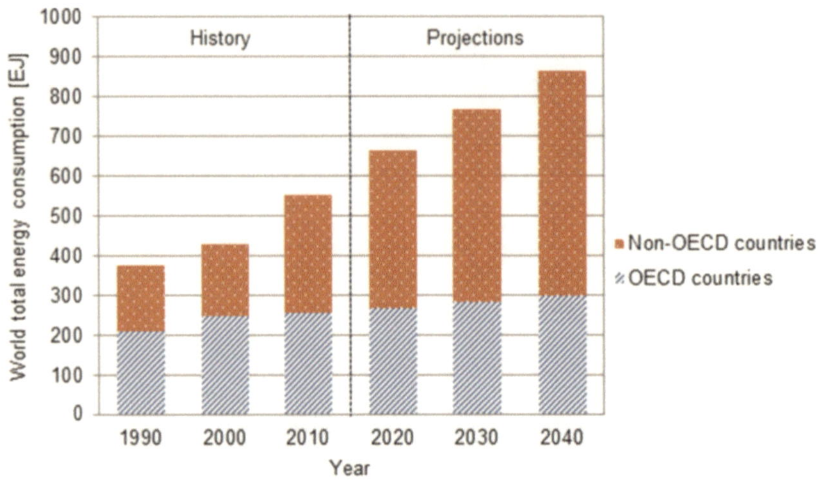

Figure 1.
World total energy consumption between 1990 and 2040 (modified from U.S. Energy Information Administration [1]). OECD, Organisation for Economic Co-operation and Development.

that, the world total energy consumption amounted 552.82 EJ in 2010 is expected to increase to 664.65 EJ in 2020 and further to 865.1 EJ in the year 2040. This corresponds to an approximately 56% increase between 2010 and 2040 [1].

Exploitation of fossil resources, like carbon and oil, for energy (e.g., electricity) generation satisfies about 70% of the global energy demand [2–5]. However, fossil resources are the main causes of greenhouse gas (GHG) emissions, which poses a detrimental effect on the environment. The combustion of fossil fuels (i.e., coal, oil, natural gas) leads also to pollution of water and land resources in the course of sulfur and nitrogen oxide emissions (e.g., acid rain) [4]. Besides pollution, these resources are also not available in unlimited quantities. The deployment of renewable energy sources, like sunlight and wind, for energy production is seen as an alternative to fossil resources. However, these energy sources are not always available, which makes it difficult to follow the variable load and meet the ever-increasing energy demand [3, 5]. Still, energy production by means of renewable energy sources is seen as a part of the future electricity grid, mostly referred to as smart grid, which will coexist as a decentralized energy source alongside with the traditional centralized power plants. The smart grid can be understood as a constant improvement of the current electricity grid. It will enable not only transport of electricity but also information, which will, on the other hand, result in a more efficient grid management, and facilitate a large-scale deployment of distributed renewable energy sources.

The realization of the smart grid, with the mentioned features, i.e., two-way information exchange in a timely manner and integration of renewable energy sources, will only be possible by a pervasive deployment of information and communication technologies (ICTs) on top of it [3]. It is the information and communication technology in the smart grid which will improve the efficiency of current electricity production, distribution, and consumption, as well as its management, and allow the integration of distributed renewable energy sources. This fact gives ICTs a very important role in smart grids, making them a very involved part of the overall electricity supply system. ICT represents the most important part in the shift from the current electricity grid to the future smart grid and will be the engine for its realization. The future electricity grid will be augmented by a magnitude of additional ICT components and devices, i.e., ICT equipment. Smart meters, power line communication (PLC) modems, data concentrators, data and control center (DCC) servers, switches, and routers are just some of them. All of these components and devices will lead to a further increase in electricity consumption, which should be taken into account in an overall, holistic analysis of environmental impacts of smart grids.

Energy efficiency is an important design parameter, and new systems should be designed with optimized energy consumption requirements in mind. Even though the operation (or use) phase of ICT equipment is important, it is only a part of the entire "story." In order to design an energy-efficient and environmentally sustainable system, other life cycle phases of ICT equipment such as raw material extraction and processing, manufacturing and assembly, recycling and disposal, as well as transportation have also to be taken into account [6]. An exergy-based life cycle assessment (E-LCA) makes such a life cycle assessment possible, as it allows an exergy consumption evaluation across the entire ICT equipment lifetime [7], which serves as a measure for the attained environmental sustainability. Exergy can be understood as the amount of energy that can be transformed into useful work, i.e., the quantity of energy available to be consumed [6]. An exergy analysis provides the means to evaluate and compare various systems with regard to their environmental sustainability. For that reason, it can be concluded that the environmental sustainability of ICT equipment relies upon its lifetime exergy (i.e., available energy) consumption and not just the electricity consumption during operation [8]. The exergy concept will be explained in more detail in the next section.

It is also worth noticing that the deployment of ICTs in various other sectors will be responsible for great emission reductions. Smart grids are just one but maybe the most promising of them [9]. Others include, e.g., smart transportation, smart infrastructure, smart production, and smart buildings. According to the Global e-Sustainability Initiative [9], ICT has the potential to enable 7.8 gigatons (Gt) of carbon dioxide equivalent (CO_2e) emission abatements by the year 2020. Smart grids will allow 2 Gt CO_2e emission abatements, which represents the strongest reduction potential of all the considered technologies. Although ICT's own footprint is expected to increase from 0.5 Gt CO_2e in 2002 to 1.4 Gt CO_2e in 2020, the enabled abatements achieved by its introduction in the different sectors will be greater. They will account for five times of ICT's own footprint, which equals to 15% of the projected total global CO_2e emissions [10]. The findings provided in the Global e-Sustainability Initiative [9] suggest that the realization of the smart grid from an environmental aspect is justified, as its potential to improve the overall environmental sustainability will overcome the environmental sustainability issues associated with the introduction of additional ICT equipment in its various domains. However, the study presented in the Global e-Sustainability Initiative [9] did not address explicitly the environmental impact of ICT for smart grids. Additionally, it used traditional LCA approaches and energy analysis. An energy analysis tracks material and energy flows of a process, enabling a complete assessment of a system [7]. Even though mass and energy conservation are included, it does not consider the second law of thermodynamics. This fact is the main drawback of an energy analysis, since different forms of energy cannot be directly compared [7].

A life cycle assessment (LCA) represents a framework for indicators that can be used to assess how various products or processes impact the environment [7]. For that purpose, all inputs and outputs of a product or process during its considered lifetime are analyzed, i.e., the evaluation takes the entire product or process life cycle under consideration. There are a lot of variants of a LCA, but most of them base their assessment on emissions. A LCA provides a thorough assessment of environmental effects but has also a few drawbacks. The most important one is that it does not produce a simple and unambiguous outcome, which could be used for easy and meaningful comparison purposes between various potential approaches. The other one is its time exposure and accomplishment expenses [7].

An exergy-based life cycle assessment (E-LCA), on the other hand, tracks the lifetime exergy consumption and considers the second law of thermodynamics. Exergy is defined as the maximum amount of useful work that can be attained from a system when brought into thermodynamic equilibrium with its reference environment [10]. Exergy can be understood as the amount of energy that can be used, i.e., the quantity of energy that can be transformed into useful work. Due to irreversibilities (i.e., inefficiencies) attributed to real processes, it is never conserved. This is the main characteristic which distinguishes exergy from energy [6]. An exergy analysis eliminates the main drawbacks of an energy analysis and a LCA. In contrast to an energy analysis, exergy analysis allows different forms of energy to be directly compared, since it makes use of the second law of thermodynamics. It does not allow a detailed assessment of environmental effects of ICTs, but it produces a simple (i.e., a single) outcome, which can be more easily computed and compared with other approaches [7]. An E-LCA is also not that time-consuming and costly to accomplish like a LCA. All of these benefits make E-LCA the best candidate for the evaluation of the environmental sustainability of ICTs for smart grids, and this thermodynamic-based indicator [7] will therefore be used as the environmental sustainability indicator of choice for the study presented in this chapter.

2. Framework for environmental sustainability analysis

This section provides the framework for the environmental sustainability analysis of information and communication technologies (ICTs). Since a large amount of additional ICT equipment is expected to become part of the future smart grid, the means to provide useful and meaningful information on the environmental sustainability of this equipment would prove beneficial. Exergy-based life cycle assessment (E-LCA) provides such means, as it allows various approaches to be compared with each other based on their exergy consumption in their different lifetime or life cycle stages, i.e., raw material extraction and processing, manufacturing and assembly, operation, recycling and disposal, as well as transportation. The obtained exergy consumption serves thereby as a measure for the attained environmental sustainability. Moreover, specific electrical generation systems and their respective energy and exergy efficiencies can be considered as well.

2.1 Classification of sustainability indicators

Before the discussion of sustainability indicators, a definition of sustainability deems appropriate. According to the Report of the World Commission on Environment and Development: Our Common Future [11, 12], the sustainable development is defined as "development that meets the needs of the present without compromising the ability of future generations to meet their own needs." However, this definition of sustainability exhibits two major shortcomings, namely the terms needs and future generations, which are not precisely specified [13]. There are also other definitions for sustainability, but they all fail to give a clear understanding if a product, system, process, or approach is sustainable or not. For that reason, the existence of a sustainability indicator, which could be used for comparison purposes between different approaches, would prove beneficial. Such an indicator could be used to evaluate which of the various approaches under consideration is the most sustainable one. With this in mind, a strict definition for sustainability would not be needed and could be replaced with a more easily attainable approach for sustainability analysis and evaluation [13].

There are several sustainability indicators in existence. Many of them have been introduced and recommended over the last years [14]. According to a report published by the Scientific Committee on Problems of the Environment (SCOPE), sustainability indicators can be categorized into three "pillars" (also called the triple bottom line), which classify sustainability into social, environmental, and economic indicators [14]. An example of a social indicator is the human development index (HDI), which evaluates the development of a country based on people and their capabilities [15]. The gross domestic product (GDP), which indicates the economic condition of a country, is an example of an economic indicator [16]. ICTs may have an influence on the entire triple bottom line of sustainability. Introduction of ICTs may lead to an improved access to education and its quality and also to more profitable markets, which will, on the other hand, result in an increase of a country's HDI (i.e., social sustainability) and the GDP (i.e., economic sustainability), respectively [14].

To obtain a useful indicator, a few requirements have to be fulfilled. First, it has to be interchangeable so that it can be modified when new data becomes available or some new processing techniques are applied. Further, the difficulty to obtain the indicator should be kept within bounds. Finally, a considerable and most desired indicator should at best provide a single value (i.e., outcome), which could be utilized to compare different approaches with each other, with the aim to obtain the most sustainable one [13].

In the following two subsections, only sustainability indicators associated with environmental effects will be considered and further discussed. After a brief discussion of environmental sustainability indicators, thermodynamic-based environmental sustainability indicators applicable to ICTs will be discussed in more detail. The most promising of them will be chosen as the environmental sustainability indicator for the assessment of ICTs in this study.

2.1.1 Environmental sustainability indicators

Environmental sustainability indicators are used to estimate the influence of human actions (i.e., their behavior) on the environment. They can be used for environmental impact assessment purposes, allowing different approaches to be compared with regard to environmental sustainability. As an example, the environmental sustainability index (ESI) represents an environmental sustainability indicator [17]. By weighing 76 different variables, a single value for a country's environmental sustainability is derived [13, 17]. The main drawback of the ESI is the fact that it is obtained based on subjective assumptions and conclusions, which lead to inaccuracies. Hence, basing the various variables of the ESI on other assumptions would most probably result in a different outcome. This leads to the conclusion that a more meaningful, unambiguous, and reliable indicator is needed, one that is based on scientifically accurate estimations, and not on vague assumptions [13].

The fundamental laws of thermodynamics, which allow assessing mass and energy transfers attributed to various processes, make such accurate estimations possible. Mass conservation and the first law of thermodynamics provide means to evaluate mass and energy transfers. The second law of thermodynamics enables a further estimation of the exploited energy, i.e., a determination of the energy being utilized [13]. It can be concluded that thermodynamic theory exhibits important advantages for an environmental sustainability analysis, e.g., evaluation of materials needed by a process and those generated due to its existence and determination of the energy demanded by the process. Moreover, it is possible to provide information on how efficiently the energy is being exploited by the process. Based on that, an evaluation of different approaches is facilitated. Even though a thermodynamic analysis may not always provide a simple value (i.e., outcome) for straightforward and uncomplicated comparison purposes, it is still possible to estimate and evaluate all inputs and outputs to and from a process, respectively. Other thermodynamic indicators, like exergy consumption, serve as a single value, i.e., a simple indicator, which allows an easy comparison between competing approaches.

In order to distinguish between various thermodynamic indicators, two main framework conditions exist [7]. The first one considers the thermodynamic quantity assessed by the indicator. Thermodynamic quantities, which can be assessed by an indicator, include, e.g., mass flow, energy flow, and exergy flow. The second parameter considers the scope of the study, which is, moreover, related to its objectives [7]. Indicators may take only a part of a device's life cycle into consideration, like the operation, manufacturing and assembly, or the recycling and disposal stage. That means, the life cycle of a device is not strictly defined. On the other hand, the cradle-to-grave approach includes raw material extraction and processing, manufacturing and assembly, operation, recycling and disposal, as well as the transportation between the various process stages. These life cycle stages can in general be seen as the most important and significant ones. The cradle-to-grave life cycle approach is commonly viewed as an entire life cycle and will therefore be adapted as the ICT component and device (i.e., ICT equipment) life cycle in this study. Thermodynamic-based environmental sustainability indicators that will be considered in the following include [9]:

- Energy analysis
- Life cycle assessment (LCA)
- Exergy-based life cycle assessment (E-LCA)

These thermodynamic-based environmental sustainability indicators will be discussed in more detail in the following subsection. A comparison between them is provided with the aim to derive the most suitable one for the environmental sustainability analysis of ICTs. **Figure 2** summarizes the discussion about sustainability indicators and depicts their classification.

2.1.2 Thermodynamic indicators suitable for sustainability analysis of ICT

In the following few parts of this subsection, energy analysis, LCA, and E-LCA will be discussed. Advantages and disadvantages of these thermodynamic indicator types are presented. In addition, basic theory behind energy, exergy, and entropy will be provided. A comparison between LCA and E-LCA for the environmental sustainability analysis of ICT equipment is presented. Further, the relation between environmental impact, exergy efficiency, and environmental sustainability is briefly studied.

2.1.2.1 Energy analysis

It is a matter of common knowledge that it is possible to store energy within systems (e.g., in batteries). Moreover, it is possible to convert energy from one form to another (e.g., coal energy to electrical energy) and to transfer it from one system to another. In the course of all the storages, conversions, and transfers, the entire quantity of energy must be conserved [18]. This fact is embodied in the first law of thermodynamics, which states that the change of the internal energy (U) of a

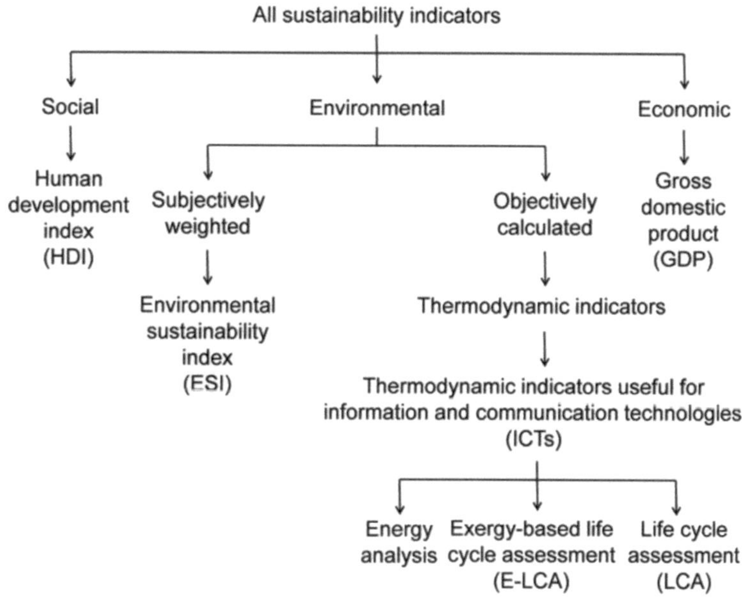

Figure 2.
Classification of sustainability indicators (modified from Lettieri et al. [7]).

system is equal to the sum of heat (Q) supplied to the system and work (W) done by the same system on its surroundings [6], i.e.

$$dU = \delta Q - \delta W. \tag{1}$$

Therefore, the energy balance of an entire process or system is equal to zero [19]:

$$En_{\text{in}} - En_{\text{out}} = 0. \tag{2}$$

According to energy balance, input and output energies of a process or system are equal. A portion of the energy at the output may be converted, e.g., into waste heat. This waste heat may as well have a positive side effect, since it could be exploited for, e.g., heating purposes. Nevertheless, the total amount of energy is conserved in all the storages, transfers, and conversions. Using an energy balance, it is possible to assess the energy demanded by a system. However, it does not tell us how well the energy is being exploited by the same system. With an energy analysis, it is not possible to determine the true thermodynamic inefficiencies related to an energy transformation system [18]. An energy analysis can identify merely the inefficiencies arising due to energy transfers out of a system that cannot be exploited anymore in the considered or some other system.

This leads to the conclusion that utilizing energy as an indicator for the assessment of energy system advantages and imperfections can be quite unclear and inaccurate [10]. An energy analysis enables quantifying energy flows. For that reason, a complete assessment of a system is facilitated. Nevertheless, even a complete life cycle energy analysis exhibits considerable disadvantages [7]. The main reason for this is the fact that an energy analysis does not consider the second law of thermodynamics. Because of that, it is not possible to directly compare distinct forms of energy with each other. However, an exergy-based life cycle assessment (E-LCA) provides means to estimate the exergy consumption over the entire considered lifetime of a system and takes the second law of thermodynamics under consideration. Based on that, it is further possible to provide information on the quality of energy, i.e., how efficiently the energy is being utilized by a system. E-LCA will be given its own discussion, and its benefits will be explained in more detail.

2.1.2.2 Life cycle assessment (LCA)

A life cycle assessment (LCA) provides means to estimate environmental effects of a product, process, or system over its respective lifetime, i.e., under consideration of its entire life cycle [20]. It was specified by the International Organization for Standardization (ISO) 14040 family of standards [14, 15]. LCA cannot be directly or precisely considered as an indicator. It rather defines a framework for indicators that can be used to evaluate how various products, processes, or systems impact the environment during their entire lifetime [13]. Based on that, various production approaches and techniques may be analyzed with regard to their environmental effects, with the aim to indicate the most efficient (i.e., environmentally sustainable) one [19]. The evaluation of environmental effects can be accomplished by capturing all inputs and outputs of a product, process, or system during its considered lifetime. Even though there are many variants of a LCA, most of them are based on emissions [13]. **Figure 3** depicts schematically the life cycle assessment (LCA) framework, which is accomplished in four phases [15]:

1. Goal and scope definition

2. Inventory analysis

3. Impact assessment

4. Interpretation

The goal and scope definition phase considers the aims of the assessment and the constraints that have to be taken into consideration. The goal definition includes

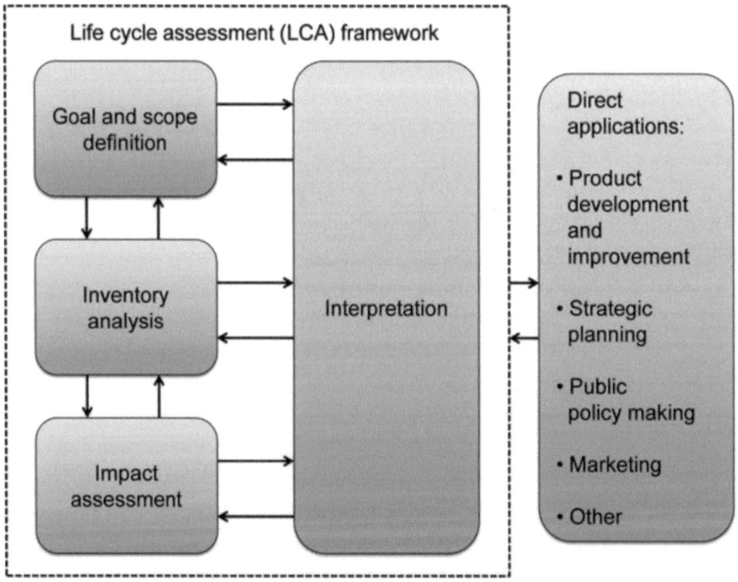

Figure 3.
Life cycle assessment (LCA) framework (modified from ISO 14040 [15]).

further the purpose of the assessment and defines the group of people potentially interested in the LCA outcomes. The scope definition defines additionally, e.g., the functional unit, data specifications, assumptions, and constraints of the assessment [15, 19]. The inventory analysis considers all inputs and outputs related to the various processes of the considered system. Inputs and outputs of the processes can be divided into economic and environmental flows [19]. Environmental inputs and outputs, which either come from, or are emitted (i.e., released) to, the environment, are considered as environmental flows. The result of the inventory analysis can be found in the inventory table. This table comprises all the resources and toxic substances that are captured or released during the considered lifetime, i.e., during the entire life cycle. Impact assessment represents the third phase of the LCA framework. During this phase the data in the inventory table is evaluated. For this purpose, impact categories are deployed, e.g., climate change, human toxicity, ozone layer depletion, and ecotoxicity [19]. In the final, i.e., interpretation, phase of a LCA, the obtained outcomes are evaluated and clarified. Based on this assessment, it is possible to suggest further improvement potentials.

A LCA provides a thorough assessment of environmental effects but has also a few drawbacks because of such a complete analysis. The first, and probably the main, disadvantage of a LCA is the lack of a simple (i.e., single) outcome which could serve as the basis for assessment and evaluation purposes between various approaches. For that reason, a direct comparison between different impact categories shows to be not that easy (e.g., ecotoxicity vs. global warming potential) [7]. Even though collections of standardized impact factors are available, the estimation of diverse environmental effects, caused from various processes, is still based on subjective assumptions and conclusions [7]. The inaccuracies introduced in the course of such vague assumptions may not be tolerable and would most probably lead to uncertain or even useless results. The second disadvantage of a LCA relates to its time exposure and accomplishment expenses. LCA software tools are existent and in use (e.g., SimaPro, GaBi Software, EarthSmart) [13, 16, 21, 22]. Still, other environmental sustainability indicators are easier to generate compared to a LCA. Exergy-based life cycle assessment (E-LCA) is an example of such an environmental sustainability indicator and will be discussed in the following part of this subsection.

2.1.2.3 Exergy-based life cycle assessment (E-LCA)

Exergy is defined as "the maximum theoretical useful work, which can be obtained from a system when brought into thermodynamic equilibrium with its environment while the system interacts with this environment only" [18]. The environment required for computing the exergy values is known as the exergy reference environment or thermodynamic environment. It is free from any irreversibility. Moreover, the exergy value of this environment must be zero. The natural environment does not meet the needs of the exergy reference environment, since it is not in equilibrium. For that reason, a model for the thermodynamic environment is always presumed. In most cases, the actual local environment is chosen as the exergy reference environment. In other words, exergy can be understood as the quantity of energy that can be utilized to perform useful work, i.e., the amount of energy available to be consumed [6]. The theory behind exergy, its characteristics, and benefits are defined by the first and second law of thermodynamics. The first law of thermodynamics defines the concept of energy conservation (see Energy Analysis). The second law of thermodynamics introduces additionally the concept of non-conservation of entropy [10, 19], i.e., the concept of entropy increase. For that reason, an exergy analysis may be utilized to evaluate, construct, and upgrade various systems. Before further discussing exergy, a few words will be addressed to entropy.

In conjunction with thermodynamic processes, entropy may be understood as the amount of energy deficiency present to perform useful work. As characterized by Rudolf Clausius, entropy can be considered as a state function of a reversible cyclic process, known as the Carnot cycle, which states that the change of entropy (S) is proportional to the heat (Q) divided by the absolute temperature (T), i.e.

$$dS = \frac{\delta Q}{T}. \tag{3}$$

It is important to underline that the process assumed in Clausius definition is a reversible one. However, irreversible processes usually lead to an increase of entropy. Assuming an isolated system is subjected to some process, entropy can only be greater or equal to zero:

$$\frac{dS}{dt} \geq 0. \tag{4}$$

This relation is satisfied with equality only for a reversible process, whereas for an irreversible process, the entropy will be greater than zero. One further aspect with regard to entropy should be mentioned. It is possible to distinguish thermodynamic entropy from logical entropy [6]. According to The American Heritage Dictionary, thermodynamic entropy (expressed in Joule per Kelvin, i.e., J/K) is formulated as "the quantitative measure of the amount of thermal energy per unit temperature not available to do work in a closed system." Logical entropy, on the other hand, can be understood as "a measure for randomness in a closed system." The entropy concept can be applied in many fields of research, e.g., thermodynamics, communications, and statistics [6]. Even though an entropy balance provides information on inefficiencies associated with a considered system based on entropy creation, it fails to communicate the precise amount of energy being exploited by the system. The entropy concept does not provide any information about the quality of energy, just like an energy analysis. However, an exergy analysis eliminates this drawback related to an energy analysis and the entropy concept. This is because exergy allows evaluating the quality of an energy carrier (i.e., its ability to perform useful work) [18].

Using an exergy analysis instead of an energy analysis, a more thorough assessment of a system can be accomplished [20]. The data obtained from an exergy analysis contains more valuable and relevant information than the one gained from an energy analysis. Further, an increase of efficiency and means to decrease thermodynamic losses may as well be realized and determined using an exergy analysis. Moreover, an exergy analysis allows different forms of energy (i.e., with different qualities) to be compared directly with each other, as the main criterion is the capability to perform useful work. For that reason, an evaluation of various systems can be accomplished in a more accurate and meaningful way. Environmental effects and improvement potentials may as well be analyzed and assessed with the use of an exergy analysis [10]. In contrast to energy, exergy is destructed and never conserved for a majority of real processes because of irreversibilities. As opposed to an energy analysis, where the energy balance for an entire process equals zero (see the paragraph Energy analysis), the exergy balance corresponds to irreversibilities related to the process under consideration [19], so that

$$Ex_{\text{in}} - Ex_{\text{out}} > 0. \tag{5}$$

According to the energy balance, the difference between input and output energy is zero, i.e., they are equal (see Eq. (2)). The exergy balance, on the other hand, defines a decrease of the quality of energy (i.e., exergy) during a process [10]. This fact is obvious from Eq. (5), according to which the exergy at the input of a system is greater than the exergy at its output. The exergy balance, which corresponds to irreversibilities associated with the considered process, is proportional to the creation of entropy (ΔS) weighted by the temperature of the exergy reference environment ($T|0$) [6, 19]:

$$Ex_{\text{loss}} = \Delta Ex = Ex_{\text{in}} - Ex_{\text{out}} = T_0 \Delta S > 0. \tag{6}$$

This relation is also known as the law of exergy loss, or the law of Gouy-Stodola [19]. The exergy loss of a complete system can be obtained by assessing and summing up the exergy losses of its corresponding subsystems or components, i.e.

$$Ex_{\text{loss, system}} = \sum Ex_{\text{loss, component}}. \tag{7}$$

Exergy losses can be divided into internal and external exergy losses. External exergy losses are composed of waste and exergies emitted (i.e., released) from a system. They encompass the amount of exergy that cannot be exploited to perform useful work. Internal exergy losses are losses related to internal inefficiencies (i.e., irreversibilities) of processes, which lead to a decrease of energy quality. They can further be classified into technical and structural exergy losses [19]. Technical exergy losses originate from system inefficiencies, while structural exergy losses are defined by various system assumptions, its composition, and features. Technical exergy losses can be minimized by applying improvement procedures. A decrease of structural exergy losses may only be achieved by modification and upgrading measures of the considered system [23].

The exergy-based life cycle assessment (E-LCA) takes all exergy inputs to a system or process during its entire lifetime into account, i.e., it includes all exergy inputs over the entire system or process life cycle [24]. Moreover, an accumulation (i.e., conservation) of exergy during the considered life cycle exergy assessment is excluded [25]. For that reason, it can be concluded that the life cycle system or process exergy losses have to be proportional to the overall exergy consumption [19, 25]. This total exergy consumption can be further used to assess and interpret the environmental sustainability of different approaches, systems, and processes.

In the following few lines, the composition of the total exergy of a system is presented. If electrical, magnetic, nuclear, and surface tension effects can be excluded, then it is possible to divide the total exergy of a system (Ex_{system}) into four components, namely, physical exergy (Ex_{physical}), chemical exergy (Ex_{chemical}), kinetic exergy (Ex_{kinetic}), and potential exergy ($Ex_{\text{potential}}$) [18], i.e.

$$Ex_{\text{system}} = Ex_{\text{physcial}} + Ex_{\text{chemical}} + Ex_{\text{kinetic}} + Ex_{\text{potential}}. \tag{8}$$

The physical exergy can be calculated according to Frangopoulos [18]:

$$Ex_{\text{physical}} = (U - U_0) + p_0(V - V_0) - T_0(S - S_0). \tag{9}$$

Here, U, V, and S are internal energy, volume, and entropy of the system, in that order. The subscript 0 in Eq. (9) indicates the state of the considered system at temperature T_0 and pressure p_0 related to the exergy reference environment. The physical exergy of a system can be further divided into thermal exergy (Ex_{thermal}) related to the system's temperature change from the temperature of the exergy reference environment and mechanical exergy ($Ex_{\text{mechanical}}$) introduced because of a difference of the system's pressure from the pressure of the exergy reference environment [18], i.e.

$$Ex_{\text{physical}} = Ex_{\text{thermal}} + Ex_{\text{mechanical}}. \tag{10}$$

In analogy to this, it is also possible to split the chemical exergy into reactive exergy (Ex_{reactive}) and nonreactive exergy ($Ex_{\text{nonreactive}}$) [18], i.e.

$$Ex_{\text{chemical}} = Ex_{\text{reactive}} + Ex_{\text{nonreactive}}. \tag{11}$$

The chemical exergy can be understood as the hypothetical maximum of useful work that can be attained from a system as this chemically equilibrates with its exergy reference environment. In order to determine the chemical exergy, it is not sufficient to define just the temperature T_0 and pressure p_0 but also the chemical consistency of the exergy reference environment. Finally, the kinetic and potential exergies can be calculated according to the following Eqs. (10):

$$Ex_{\text{kinetic}} = \frac{mv^2}{2} \tag{12}$$

$$Ex_{\text{potential}} = mgh. \tag{13}$$

As can be seen from Eqs. (12) and (13), kinetic and potential exergies are equal to kinetic and potential energies. The variables v and h correspond to the velocity and height relative to that of the exergy reference environment ($v_0 = 0, h_0 = 0$) [18].

The preceding two parts of this subsection elaborated in detail two important but different approaches for an environmental sustainability analysis of ICTs, namely, the life cycle assessment (LCA) and the exergy-based life cycle assessment (E-LCA). In the following part of this subsection, a comparison between them is provided with the aim to point out the advantages of E-LCA for the assessment and evaluation of environmental effects, i.e., the environmental sustainability, associated with various ICT equipment.

2.1.2.4 LCA vs. E-LCA

A life cycle assessment (LCA) enables a thorough assessment of environmental effects related to ICT equipment, by capturing all inputs and outputs during its considered lifetime [26–28]. Most LCA approaches base their analysis of environmental effects on emissions, which pose a potential to negatively impact the environment, e.g., greenhouse gas (GHG) emissions. However, such an analysis brings also considerable drawbacks with it, like time exposure, accomplishment expenses, and, more importantly, the absence of a simple and unambiguous outcome for meaningful comparison purposes between different approaches. The exergy-based life cycle assessment (E-LCA), on the other hand, represents a sound approach for the environmental sustainability analysis of ICT equipment, based on its lifetime exergy consumption, which serves as a measure for the attained environmental sustainability. The major benefit of E-LCA is the fact that it leads to a single outcome (i.e., an exergy consumption value) that can be easily compared with various other potential approaches. Moreover, the quite moderate time exposure, accomplishment expenses, as well as the update and expandability features of E-LCA make this thermodynamically based indicator the best choice for the environmental sustainability analysis of ICTs.

The conclusions drawn from the E-LCA do not differ much from those of a LCA. Actually, they coincide quite well in all life cycle stages of the considered ICT equipment, the only difference being the relative scales and the quantity characterizing the environmental effects (e.g., GHG emissions vs. exergy consumption). **Figure 4** shows the results of a LCA of three different smartphones, namely, an Apple iPhone 4S, a Nokia Lumia 920, and a Huawei U8652 [29]. The contribution of the different smartphone life cycle stages to the climate change, attributed to GHG emissions, over a lifetime of 3 years, is expressed in kilograms of carbon dioxide equivalents (kg CO_2e) [29, 30]. The production stage accounts for raw material extraction and processing, manufacturing and assembly, as well as the transportation during these two processing phases; the use phase is based on a 3-year operation of the smartphones; the transportation stage includes the transport of the products to their distribution location, while the recycling stage includes the transport of products to recycling plants, their separation, and shredding. It is evident that the use of a LCA for the assessment of environmental effects, even for the three considered smartphones, leads to relatively different outcomes (ranging between 16 and 70 kg CO_2e), which is a result of the quite different assumptions made for each of these approaches and LCA tools used for their assessment. Even this simple example illustrates the underlying ambiguity of a LCA. However, the conclusions drawn from each of these approaches suggest that the most dominant life cycle stage of a smartphone relates to its production stage, i.e., its raw material extraction and processing, as well as its manufacturing and assembly (including transport).

The same conclusions on the environmental sustainability are provided by means of the E-LCA. **Figure 5** depicts the LCA and three E-LCA use cases (UCs) of an Apple iPhone 5C, respectively [31]. The assumptions provided in Ref. [31], i.e., those concerning the production, transportation, customer use, as well as recycling, have been adapted to a good degree into the E-LCA framework, in order to obtain a more meaningful comparison between these two environmental sustainability analysis approaches. The three depicted E-LCA UCs assume different usage intensities and therefore different daily smartphone charging durations. The first UC assumes a daily charging duration of 4 h, the second UC of 8 h, and the third UC of 12 h. An operational duration of 3 years is assumed. It should be noted that the environmental impact of the different smartphone life cycle stages in **Figure 5** is given in percentage of the lifetime GHG emissions in the case of the LCA and in percentage of the lifetime exergy consumption in the case of the E-LCA for easier comparison purposes. **Figure 5** suggests that for both the LCA and the E-LCA approach, the most dominant (i.e., environmentally relevant) life cycle stage of the

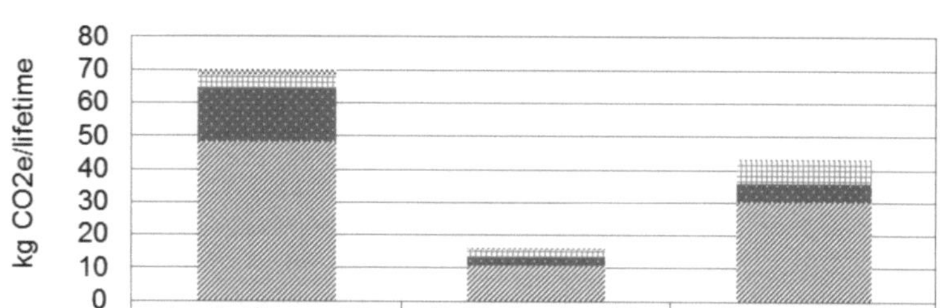

	Apple iPhone 4S	Nokia Lumia 920	Huawei U8652
Recycling	1,4	0,2	0,1
Transportation	4,2	2,4	6,9
Use	16,1	2,7	5,9
Production	48,3	10,7	30

Figure 4.
Life cycle assessment (LCA) of three different smartphones (modified from Andrae and Vaija [29]).

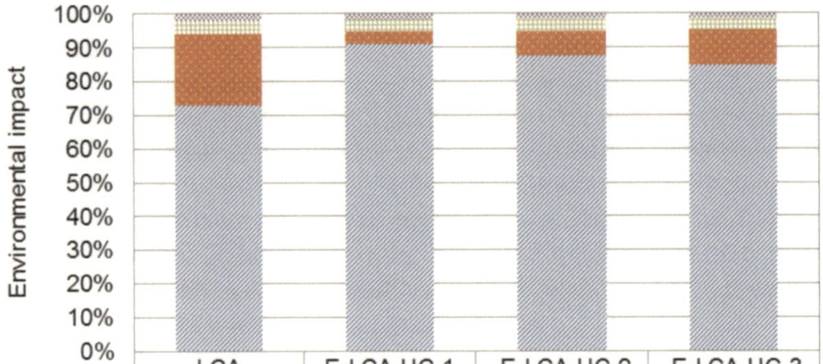

	LCA	E-LCA UC 1	E-LCA UC 2	E-LCA UC 3
Recycling	2,00%	1,59%	1,54%	1,48%
Transportation	4,00%	3,74%	3,61%	3,48%
Use	21,00%	3,74%	7,22%	10,45%
Production	73,00%	90,92%	87,64%	84,59%

Figure 5.
Life cycle assessment (LCA) and three exergy-based life cycle assessment (E-LCA) use cases (UCs) of an Apple iPhone 5C.

smartphone relates to its production. The contribution of the use phase to environmental effects is more pronounced in the case of the LCA than in the case of the E-LCA. This is mainly related to the power grid mix assumed in Ref. [31] (which is accounted at a continent level), as well as the deployed LCA tool. Such an approach underlines again the ambiguity of a LCA, as different power grid mix assumptions and LCA tools will unavoidably lead to different relative outcomes (compare also the use phases of the three different smartphones depicted in **Figure 4**). E-LCA, on the other hand, bases its assessment of the use phase on the operational exergy consumption, leading to a more scientifically reasonable and justified approach. The difference between the use phases of the LCA and E-LCA becomes smaller assuming a longer daily charging duration (i.e., usage intensity).

For that reason, it can be argued that the conclusions drawn from a LCA and the E-LCA coincide to a good degree with each other. The main benefit of E-LCA, however, is the fact that it leads to a simple and unambiguous outcome, useful for easy and meaningful comparison purposes between various potential approaches. Such a feature lacks in a LCA, as each environmental sustainability assessment approach includes different assumptions, as well as LCA tools, leading to relatively different results (see **Figure 4**). Therefore, E-LCA is chosen as the environmental sustainability indicator of choice for the assessment of ICT equipment in the scope of this study.

To summarize the discussion regarding the thermodynamically based environmental sustainability indicators that can be applied to ICTs, **Table 1** provides a comparison between the discussed indicator types. It can be concluded that the environmental sustainability of ICTs may at best be assessed and evaluated using the E-LCA approach. E-LCA proved to be the best choice among the other discussed thermodynamically based environmental sustainability indicators, namely energy analysis and LCA. It allows assessing the lifetime exergy consumption of various ICT approaches, which will, on the other hand, serve as an indicator for the attained environmental sustainability. Even though E-LCA does not provide detailed information of environmental effects like a LCA, it derives at a simple outcome (i.e., a single value) that can be used to compare different approaches, systems, and processes. If more information of a particular system is needed, and a more thorough assessment deems appropriate, a LCA can be applied [7]. Moreover, E-LCA satisfies the requirements defined by SCOPE for useful and applicable indicators.

Indicator type	Advantage	Disadvantage
Energy analysis	Enables energy assessment and evaluation by the use of the first law of thermodynamics	Different forms of energy cannot be directly compared; environmental effects cannot be directly assessed
Life cycle assessment (LCA)	Allows a very detailed and thorough assessment of environmental effects	Difficult to derive; lack of a simple and unambiguous outcome for easy comparison purposes
Exergy-based life cycle assessment (E-LCA)	Different forms of energy can be directly compared; simpler to obtain than a LCA; leads to a single value for easy comparison purposes	Does not allow a direct, i.e., a thorough, assessment of environmental effects

Table 1.
Advantages and disadvantages of thermodynamic indicators [9, 14].

E-LCA has the major benefit of being "flexible," as it can be modified and changed when, e.g., manufacturing techniques are changed. The new data can be quite easily incorporated into the existing assessment framework. Exergy analysis can be used to assess and analyze heterogeneous systems. Since recently, E-LCA has also been used to analyze and evaluate the environmental sustainability of ICTs [7, 8, 13, 25, 32, 33].

As the final topic in this section, the relation between environmental impact, exergy efficiency, and environmental sustainability will be discussed. There are a few ways to define exergy efficiency. One of them, known as the simple efficiency, is given by the exergy at the output divided by the exergy at the input of a system or process [19], i.e.

$$\eta_{\text{ex, simple}} = \frac{Ex_{\text{out}}}{Ex_{\text{in}}} = 1 - \frac{\Delta Ex}{Ex_{\text{in}}} = 1 - \frac{Ex_{\text{loss}}}{Ex_{\text{in}}}. \tag{14}$$

There are also other definitions for the exergy efficiency. However, the simple exergy efficiency defined by Eq. (14) serves quite well for elaboration and assessment purposes in the scope of this study. It can be argued that an increase of exergy efficiency, i.e., a decrease of exergy losses (see also Eq. (6)), is an important step toward the improvement of the environmental sustainability of a system, process, or approach. Moreover, such an improvement would most probably lead to a reduction of the environmental impact associated with the considered system,

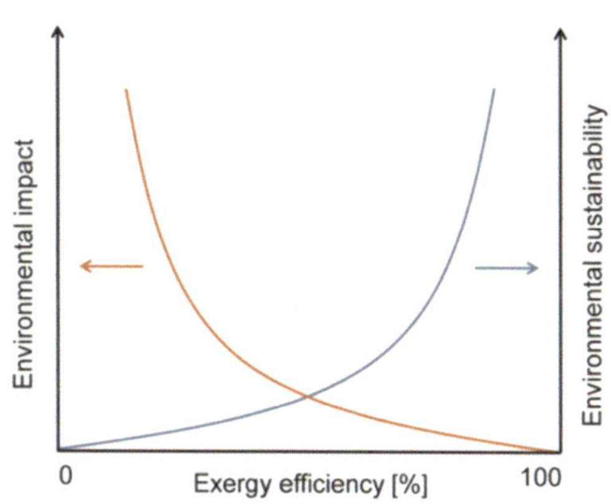

Figure 6.
Relation between environmental impact, exergy efficiency, and environmental sustainability (modified from Rosen et al. [10]).

process, or approach. **Figure 6** illustrates the relation between environmental impact, exergy efficiency, and environmental sustainability. An increase of the exergy efficiency is accompanied by a decrease of the environmental impact and at the same time by an increase of the environmental sustainability. Therefore, measures to increase the exergy efficiency, or equivalently decrease the exergy losses (i.e., exergy consumption), are desirable to achieve more environmentally sustainable systems or processes. This will, moreover, result in a decrease of environmental effects.

Even though E-LCA does not provide detailed information on environmental effects in comparison to a LCA, it can be argued that a low-exergy consumption (i.e., low-exergy losses or equivalently a high-exergy efficiency) leads to less environmental effects and, more importantly, to an increase of the environmental sustainability of the considered system, product, or approach. This observation will serve as the basis for the assessment and evaluation of the attained environmental sustainability of ICT equipment deployed in the various smart grid domains. It has to be mentioned that the relation between environmental impact, exergy efficiency, and environmental sustainability depicted in **Figure 6** does not hold for all processes. For instance, if a process deploys specific pollution control methods, it is possible to observe a decrease of environmental effects, even in the course of a reduction of the exergy efficiency, i.e., an increase of exergy losses or equivalently an increase of the exergy consumption.

2.2 Exergy consumption in different life cycle stages

In this section, the exergy consumption values for various processes (needed to estimate and analyze the environmental sustainability of ICT equipment deployed in the different smart grid domains) will be presented. Furthermore, the framework required for the evaluation of the operational exergy consumption of ICT equipment will be provided. As already mentioned in Section 1, a lot of additional ICT equipment will become part of the future electricity grid, the smart grid. The environmental sustainability of this ICT equipment will be analyzed using the exergy-based life cycle assessment (E-LCA), as this indicator type proved to be the most suitable one for the assessment and evaluation of the environmental sustainability of ICTs.

Life cycle stages that will be considered in the environmental sustainability analysis of ICT equipment include raw material extraction and processing, manufacturing and assembly, operation, recycling and disposal, as well as the transportation between the different process stages. The exergy consumed in all of these life cycle stages will be estimated and will serve as an indicator for the attained environmental sustainability of various systems and approaches. The research focus of most studies on energy efficiency and environmental sustainability of ICT equipment considers mainly their use phase, i.e., their respective power consumption during operation [8]. However, such an approach takes only a portion of the entire ICT equipment life cycle into account, since the majority of its lifetime is not included in the estimation of the overall exergy, i.e., useable energy, consumption. In order to obtain a more thorough environmental sustainability assessment, the whole life cycle of ICT equipment needs to be considered, i.e., from cradle to grave. This approach is schematically depicted in **Figure 7**, which shows the different lifetime exergy consumption stages that will be taken into account for the environmental sustainability assessment of ICT equipment deployed in the various domains of the overall model.

For this purpose, the considered overall model will be divided into submodels for the home area network (HAN)/building area network (BAN), neighborhood

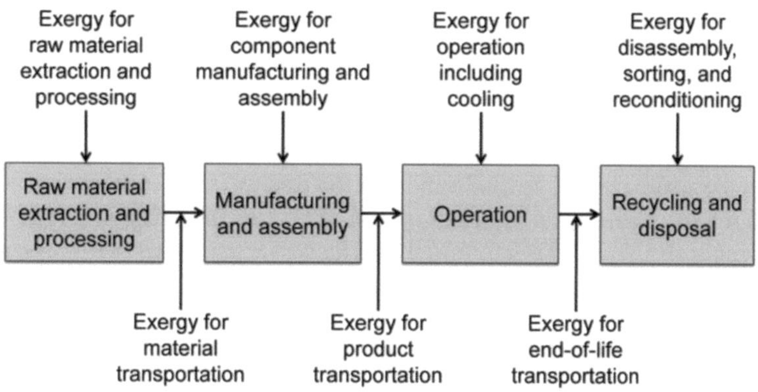

Figure 7.
Life cycle exergy consumption stages of ICT equipment (modified from Aleksic [6]).

area network (NAN), access network (AN), core network (CN), and finally the data and control center (DCC). ICT equipment deployed in the HAN/BAN includes smart meters and power line communication (PLC) modems, as well as user devices (UDs) like smartphones, tablets, notebooks, digital subscriber line (DSL) modems, and home energy management systems (HEMSs). This ICT equipment will enable various monitoring and control functions and allow an easy and efficient management of customers' electricity consumption.

A few HANs/BANs are considered a NAN. Equipment in the NAN includes, additionally to the ICT equipment of HANs/BANs, also data concentrators. Smart meters, PLC modems, and data concentrators, referred here to as utility equipment (UE), will be responsible for the collection of various quantities (e.g., electricity consumption data), their management, processing, and forwarding to the DCC where they will be further analyzed. This will enable an efficient management and control of the electricity grid. But also other benefits for the user will emerge, e.g., diverse demand response (DR) methods [3]. The Global System for Mobile Communications (GSM) and the Universal Mobile Telecommunications System (UMTS) radio access network (RAN) compose the considered AN. Components of the AN (more precisely the RAN) include base transceiver station (BTS) and base station controller (BSC) racks for the GSM RAN and Node B and radio network controller (RNC) racks for the UMTS RAN. Further, for a more complete and accurate assessment, optical fiber cables and cat5e cables will be included in the evaluation of the RAN. The CN includes serving General Packet Radio Service (GPRS) switching node (SGSN) and gateway GPRS switching node (GGSN) racks. Cables will be included in the CN as well. The DCC comprises not only a number of servers (including cooling exergy consumption), switches, routers, modems, and cables, but also notebooks, tablets, and smartphones, which represent an important part of the control center (CC), will be included as well. The total exergy consumption of the ICT equipment deployed in these various submodels will be estimated separately. Moreover, the exergy consumption in all the different life cycle stages of this ICT equipment will be evaluated, in order to see where the largest exergy consumption occurs, i.e., in which life cycle stage. Finally, the total exergy consumption of the overall system will be estimated and further analyzed. Additionally, the most dominant ICT equipment category groups (i.e., those with the largest exergy consumption), as well as the most important smart grid domains (i.e., those closely associated with environmental sustainability issues) will be indicated.

Here, a distinction will be made between embodied and operational exergy consumption. Embodied exergy consumption (EEC) refers to the exergy consumed during raw material extraction and processing, manufacturing and assembly, recycling and disposal, as well as the transportation between the various process stages. Operational exergy consumption (OEC) refers to the power consumed by the considered system during operation, called operational power consumption, and the power needed by the cooling infrastructure (if present), called cooling exergy consumption [6]. This distinction of the exergy consumption will prove beneficial for the assessment of the environmental sustainability of ICT equipment, its deployment in the various submodels of the smart grid, and finally the environmental sustainability assessment of the overall system.

2.2.1 Requirements on data for environmental sustainability analysis

Before we can start with the environmental sustainability assessment of ICT equipment, the following data has to be made available [13]:

- Mass of materials that compose the various ICT components and devices
- Exergy consumption values for extraction and processing of various raw materials
- Amount and dimensions of printed circuit boards (PCBs), integrated circuits (ICs), and processors
- Exergy consumption values for various manufacturing and assembly processes
- Operational specifications of the considered system (i.e., ICT equipment), such as peak power consumption, average load, uptime (i.e., daily operation time), operational (i.e., use) duration, and cooling characteristics
- Exergy consumption value for recycling and disposal processes of ICT equipment
- Locations of raw material extraction and processing, manufacturing and assembly, operation, recycling and disposal, and the transportation mode between these locations
- Exergy consumption values for various transportation modes

In the following few parts of this subsection, the exergy consumption values required for the ICT equipment exergy consumption (i.e., environmental sustainability) estimation in the different process stages as well as the framework for the calculation of the operational exergy consumption, will be presented and discussed in more detail. These exergy consumption values as well as the operational exergy consumption framework form the basis for the assessment of the environmental sustainability of the ICT equipment.

2.2.1.1 Raw material extraction and processing

Before going into the examination of the various exergy consumption values, it will be repeated what is understood under exergy consumption. Exergy can be understood as the amount of energy available to perform useful work. For example, high-concentration ores possess an exergy value in comparison to the Earth's crust [13]. The extraction of these ores for manufacturing and assembly (i.e., production)

DOI: http://dx.doi.org/10.5772/intechopen.88515

purposes is related to an exergy loss from the environment, i.e., exergy consumption. The estimation of the raw material extraction and processing exergy consumption of ICT equipment is achieved by using mass-specific exergy consumption values for different materials obtained from Refs. [8, 13]. For that reason, the mass of various materials, which make up the different components and devices, needs to be provided. The mass-specific exergy consumption values for raw material extraction and processing of different materials are given in **Table 2**.

The energy required for mining, transporting, and refining defines, among others, these raw material extraction and processing exergy consumption values. The exergy required for the extraction of materials or ores with high concentrations, compared to that of the Earth's crust, is also included in this process stage [13]. For a lot of materials, characterized by a minor weight, mass-specific exergy consumption values were not available. For that reason, an order of magnitude estimate approach deemed appropriate and was applied to those materials in order not to magnify their share to the overall exergy consumption in this particular life cycle stage [6]. Because of the low weight of many of these materials, the deviation from the true or exact raw material extraction and processing exergy consumption value is assumed to be very low, i.e., negligible [8].

2.2.1.2 Manufacturing and assembly

The manufacturing and assembly exergy consumption is composed of the energy required by the machinery and procedures for manufacturing and assembly purposes and the exergy contained in the resulting material waste streams. It is estimated that the waste stream for metals and plastics corresponds to 10 and 50%, respectively [13]. Mass-specific exergy consumption values for the manufacturing and assembly of metals and plastics are taken from Refs. [8, 13]. The manufacturing and assembly of printed circuit boards (PCBs), integrated circuits (ICs), and processors involves very complex and more profound energy-related techniques and procedures than those applied to metals and plastics. The exergy consumption values for the manufacturing and assembly procedures of these components are as well taken from Refs. [8, 13]. The exergy consumption expended on the manufacturing and assembly of PCBs is determined and provided on a per area basis, which is based on an average dimension assumption of PCB. The manufacturing and assembly exergy consumption of ICs is determined and provided on a per IC basis, which is based on an average dimension assumption of ICs. The manufacturing and assembly of processors relies upon highly purified silicon wafers and is accompanied by large amounts of water and chemicals. This leads further to various side product waste streams and explains, moreover, the high

Material	Specific exergy (kJ/kg)
Aluminum	341,500
Steel	52,100
Plastic	92,300
Copper	67,000
Iron	51,040
Glass	33,400
Epoxy, ceramics, and others	20,000

Table 2.
Mass-specific exergy consumption values for raw material extraction and processing [6, 14].

quantity of this exergy consumption value. The overall manufacturing and assembly exergy consumption of a system, component, or device is composed of the exergy consumption portions of the various processes involved in its production. The exergy consumption values for the different manufacturing and assembly processes can be found in **Table 3**.

As can be seen from **Table 3**, the most complex manufacturing and assembly procedures (i.e., processes with the highest exergy consumption expenditure) are those for PCBs, ICs, and processors. For that reason, the estimation of the manufacturing and assembly exergy consumption of metals and plastics can be neglected for devices with a low mass, e.g., smartphones, tablets, power line communication (PLC) modems, and data concentrators, since the portion of the more complex processes is expected to dominate.

As these devices are, furthermore, considered to be equipped with quite a few ICs and processors (i.e., higher-exergy consumption-related components), the deviation from the true or exact exergy consumption in this particular process stage is expected to be very low.

2.2.1.3 Operation

The operational exergy consumption is composed of the operational power consumption, i.e., the electricity required to power the various ICT equipment, and the cooling exergy consumption, i.e., the electricity demanded by the cooling infrastructure, if present. The framework for the evaluation of the operational power consumption and the cooling exergy consumption will be discussed in more detail in the following two segments.

1. Power consumption of the networking equipment

The operational power consumption takes only the electric power (i.e., electricity) demanded by the system (i.e., some ICT equipment) into consideration, including the cooling requirements within the considered system, i.e., internal fans. The operational power consumption of a system, expressed in joule (J), can be calculated according to the following relation (modified from Hannemann et al. [8]):

$$Ex_{\text{operational}} = P_{\text{system, peak}} \bullet \acute{L}_{\text{system}} \bullet t_{\text{up}} \bullet t_{\text{operational}} \bullet C, \tag{15}$$

where $P_{\text{system, peak}}$, given in watts (W), denotes the system's peak electricity consumption. $\acute{L}_{\text{system}}$ represents the average load of the system during its use. It is expressed in % of the peak system load. t_{up} is the system's daily operation time and is expressed in % of the time it is deployed. $t_{\text{operational}}$ denotes the system's total

Material/component (unit)	Specific exergy
Metals (kJ/kg)	0.28
Plastics (kJ/kg)	14.9
PCBs (kJ/m^2)	238,400
ICs (kJ/IC)	12,500
Processors (kJ/processor)	1,242,000

PCBs, printed circuit boards; ICs, integrated circuits.

Table 3.
Exergy consumption values for manufacturing and assembly [8, 13].

usage time, expressed in years. Finally, C represents a constant required for a correct unit conversion. It is important to mention that the operational power consumption, described by Eq. (15), takes the entire exergy that enters the considered system during its use into account. It does not depict its actual destruction. Even though the heat emitted from the system may be exploited to perform useful work (e.g., for heating purposes), most systems, however, discard this emitted heat. For that reason, Eq. (15) defines the total exergy loss (i.e., exergy consumption), which is related to the electricity consumption of the considered system.

2. Power consumption of the equipment for cooling

The cooling exergy consumption may represent a major part of the operational exergy consumption. It is important for the analysis of data centers, which house a large number of servers and other equipment (e.g., switches, routers). Its power consumption is defined by the electricity demanded by, e.g., the computer room air conditioning (CRAC) units, in general cooling equipment required for a data center's proper operation. According to Hannemann et al. [8], the cooling peak power consumption can be considered to be approximately proportional to the server's peak power consumption. The cooling exergy consumption, expressed in joules (J), can be calculated using the following relation (modified from Hannemann et al. [8]):

$$Ex_{\text{cooling}} = P_{\text{CRAC, peak}} \bullet \left[1 - \delta_{\text{dynamic}} \bullet \left(1 - \acute{L}_{\text{CRAC}}\right)\right] \bullet t_{\text{up}} \bullet t_{\text{operational}} \bullet C. \qquad (16)$$

$P_{\text{CRAC, peak}}$, given in Watt (W), denotes the CRAC units' peak power consumption. δ_{dynamic} denotes a binary indicator, which is used to indicate if a linear adjustment of the CRAC units' electricity consumption to that of the servers is in force, i.e., active. If it is, it is equal to one (i.e., $\delta_{\text{dynamic}} = 1$); otherwise it is zero (i.e., $\delta_{\text{dynamic}} = 0$). $\acute{L}_{\text{CRAC}}$ represents the average CRAC load. It is expressed in % of the peak CRAC load. This factor is needed just in the case of a dynamic cooling system (i.e., for $\delta_{\text{dynamic}} = 1$), where an adjustment of the cooling electricity consumption to that of the servers is in force. t_{up} corresponds to the CRAC units' daily operation time and is expressed in % of the time it is deployed. $t_{\text{operational}}$ denotes the CRAC units' total usage time, expressed in years. C also denotes here a constant needed for a correct unit conversion [6]. Equation (16) considers also here, equivalently to Eq. (15), the entire exergy that enters the considered cooling system (e.g., CRAC units) during its use, not its actual destruction. Even though the exergy contained in the CRAC unit waste emissions may be exploited to perform useful work, this amount is considered to be negligible and will not be considered in further assessments [6].

2.2.1.4 Recycling and disposal

It is not an easy task to obtain the exact amount of exergy consumed during the recycling and disposal (i.e., dismantling, shredding, separating, and recovering useful materials) of a particular ICT device. The main reason for this is the fact that accurate and trustworthy information on the goods (e.g., notebooks, smartphones) delivered to the recycling plants is not available. In Hannemann et al. [8], an order of magnitude estimate approach was assumed, according to which the amount of exergy expended on the recycling of a server (relative to its mass) corresponds to approximately 520 kilo joules per kilogram (kJ/kg). This value is assumed to be reasonable and will, therefore, be used in this correspondence as the reference for the evaluation of the recycling and disposal exergy consumption of ICT equipment.

2.2.1.5 Transportation

In order to make the exergy-based life cycle assessment (E-LCA) complete, the exergy consumed during the transportation between the various life cycle stages needs to be included in the analysis as well. After raw material extraction and processing, the materials have to be transported to the manufacturing and assembly location. From there, the final products will be transported to the location where they will be operated (or used). The final stage is the transportation of used up, damaged, or outdated components and devices to recycling plants. Three different transportation stages in the life cycle of ICT equipment will be considered, and these are material transportation, product transportation, and end-of-life transportation to recycling plants. The transportation exergy consumption does not only depend merely on the mass of the materials but also on the distance between the different process stages and the transportation mode [13]. **Table 4** shows the mass (and distance)-specific exergy consumption values for the different transportation modes.

One further important fact, which concerns the provided exergy consumption values, has to be pointed out. A lot of manufactured and assembled (i.e., produced) ICT components and devices, for which diverse raw materials (e.g., aluminum, steel, plastic, copper; see also **Table 2**) need to be extracted and processed, will not be used for smart grid applications only. Therefore, the total estimated raw material extraction and processing exergy consumption are weighted by a usage factor (UF) of the considered ICT component or device. This is done with the aim not to overestimate the impact of the considered ICT component or device to the total exergy consumption in this particular life cycle stage. For the same reason, the estimated exergy consumption in the other process stages (i.e., life cycle stages) needs to be weighted by such a UF as well, including manufacturing and assembly, operation, recycling and disposal, as well as transportation. User devices (UDs) represent such devices, since smartphones, tablets, and notebooks will be used for other applications as well. In fact, they will be used for other purposes most of the time (e.g., various Internet services, telephony, mobile and video games). They will be used for home energy management purposes, e.g., in average only 2 h daily. To account for this fact, the consumption of smartphones, tablets, and notebooks will be multiplied by their respective UF, referred to as home energy management usage factor (HEMUF). The UF of home energy management systems (HEMSs), on the other hand, is accounted for through its daily uptime. Other examples of ICT equipment in the smart grid, which will be used for only a few minutes daily, are the Global System for Mobile Communications (GSM) and Universal Mobile Telecommunications System (UMTS) radio access network (RAN) components, i.e., base transceiver station (BTS), base station controller (BSC), Node B, and radio network controller (RNC) racks, as well as cables connecting these racks. These components will be multiplied by their respective UF as well, termed smart grid application usage factor (SGAUF), accounting for the time they are used for the

Mode of transportation	**Specific exergy (kJ/kg km)**
Air	22.41
Truck	2.096
Rail	0.253
Ship	0.296

Table 4.
Mass (and distance)-specific exergy consumption values for different transportation modes [8, 13].

advanced metering infrastructure (AMI) smart grid application (AMI will be discussed in Section 3.1.1 in more detail). The same UF will be used for core network (CN) components, i.e., serving General Packet Radio Service (GPRS) support node (SGSN) and gateway GPRS support node (GGSN) racks, core switches, as well as cables connecting this equipment. Utility equipment (UE), i.e., smart meters, power line communication (PLC) modems, and data concentrators, will be used for the AMI smart grid application only, so their respective UF is one. The same holds for the data and control center (DCC) equipment (i.e., UF = 1), which manages and controls the entire electric power grid.

3. Sustainability analysis of ICT for smart grids

This section presents the environmental sustainability analysis of the overall system. The exergy-based life cycle assessment (E-LCA) is used as the environmental sustainability indicator of choice for the assessment and evaluation of information and communication technology (ICT) equipment, crucial for a proper operation of the advanced metering infrastructure (AMI) and home area networks (HANs). For that purpose, the overall model, developed for the environmental sustainability analysis of ICTs for smart grids, will be divided into submodels for the home area network (HAN)/building area network (BAN), neighborhood area network (NAN), radio access network (RAN), core network (CN), and the data and control center (DCC). The exergy consumption of ICT equipment deployed in the various submodels will be estimated and analyzed using the E-LCA framework. Moreover, the considered ICT equipment will be categorized into five different category groups, namely, utility equipment (UE), user devices (UDs), radio access network (RAN), core network (CN), as well as data and control center (DCC) equipment. Such an approach provides the means to indicate the most exergy consumption-related ICT equipment categories, as well as the most dominant domains of the smart grid. Based on that, ICT equipment categories and smart grid domains closely associated with environmental sustainability issues can be indicated.

3.1 Description of the overall system

We present first an ICT equipment inventory required for the E-LCA of the overall system. The considered ICT equipment defines the basis for a correct and reliable functioning of the advanced metering infrastructure (AMI) and home area networks (HANs). Further, the submodels for the home area network (HAN)/ building area network (BAN), neighborhood area network (NAN), radio access radio network (RAN), core network (CN), and the data and control center (DCC), as well as the ICT equipment included in these submodels, are provided. Finally, the assumptions and models required for the environmental sustainability analysis of ICT equipment involved in AMI and HANs are outlined.

The overall model, developed for the assessment of ICT equipment involved in AMI and HANs, is composed of submodels for the HAN/BAN, NAN, RAN, CN, and the DCC. This overall system is schematically depicted in **Figure 8**. The HAN/BAN equipment includes smart meters and power line communication (PLC) modems, as well as user devices (UDs) like smartphones, tablets, notebooks, digital subscriber line (DSL) modems, and home energy management systems (HEMSs), required for a proper utilization of the HAN application. The HEMS in **Figure 8** is placed out of the home, as it is assumed that the HEMS can support energy management requirements of a certain number of households (more precisely

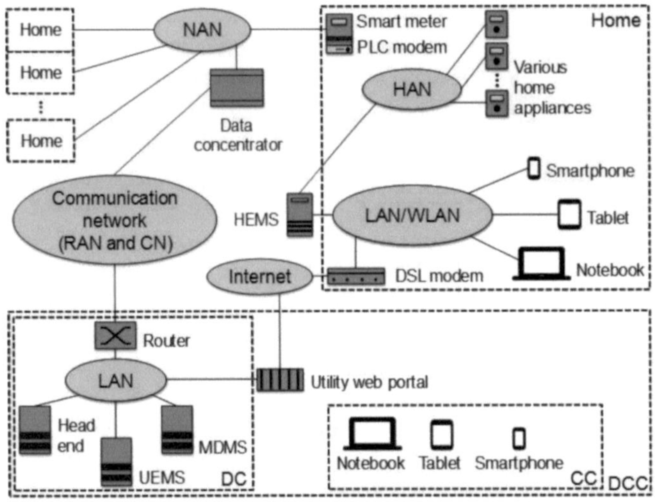

PLC:	power line communication
HAN:	home area network
LAN:	local area network
WLAN:	wireless local area network
DSL:	digital subscriber line
HEMS:	home energy management system
NAN:	neighborhood area network
RAN:	radio access network
CN:	core network
MDMS:	meter data management system
UEMS:	utility energy management system
DC:	data center
CC:	control center
DCC:	data and control center

Figure 8.
Overall model considered for the environmental sustainability analysis of information and communication technology (ICT) equipment for smart grids (modified from Budka et al. [34]).

between 10 and 100). NAN equipment includes additionally to those of the HAN/BAN also data concentrators, required for data collection, processing, and forwarding purposes, and represents a very essential component of the smart grid. RAN equipment includes the Global System for Mobile Communications (GSM) and Universal Mobile Telecommunications System (UMTS) equipment, i.e., base transceiver station (BTS), base station controller (BSC), Node B, and radio network controller (RNC) racks. The CN comprises serving General Packet Radio Service (GPRS) support node (SGSN) and gateway GPRS support node (GGSN) racks, as well as core switches. Copper and optical fiber cables, required to link RAN and CN equipment, are included in the overall system as well. The DCC comprises not only servers (including cooling exergy consumption), switches (i.e., core, aggregation, and rack/edge switches), routers, modems, and cables but also notebooks, tablets, and smartphones, which represent an important part of the control center (CC). Such a holistic approach allows us to obtain more meaningful conclusions on the environmental sustainability of ICT equipment associated with the advanced metering infrastructure (AMI) and home area networks (HANs). Lifetime assumptions of the ICT equipment deployed in the RAN, CN, as well as the DCC are provided in **Table 5**.

These lifetime characteristics are assumed to be fixed during the E-LCA of the overall system. The ICT equipment lifetime listed in **Table 5** is, moreover, weighted by its respective usage factor (UF), termed smart grid application usage factor (SGAUF), accounting for the time it is used for the AMI and HAN smart grid applications. The listed ICT equipment lifetime assumptions are partly based on analytical conclusions as well as the information provided in Refs. [32, 35]. Lifetime assumptions as well as various other parameter assumptions of utility equipment (UE) and user devices (UDs) will be provided in the respective scenario considered.

3.1.1 Models for AMI and HANs

The advanced metering infrastructure (AMI) represents the basic infrastructure of the future smart grid. It includes smart meters, PLC modems, data concentrators, and DCC equipment, as well as communication network equipment, i.e., RAN and CN equipment (see **Figure 8**). It is assumed that the smart meter measurements are

ICT equipment	ICT equipment category group	Lifetime (years)
BTS rack	RAN	7
BSC rack	RAN	8
Node B rack	RAN	8
RNC rack	RAN	9
SGSN rack	CN	10
GGSN rack	CN	10
Core switch	CN/DCC	3
Aggregation switch	DCC	3
Rack/edge switch	DCC	3
Server	DCC	4
Notebook (15-inch)	DCC	3
Notebook (13-inch)	DCC	3
Tablet	DCC	2
Smartphone	DCC	2
Router	DCC	3
DSL modem	DCC	3
Cat5e cable	RAN/CN/DCC	>20
Optical fiber cable	RAN/CN/DCC	>20

BTS, base transceiver station; BSC, base station controller; Node B, Universal Mobile Telecommunications System (UMTS) base station; RNC, radio network controller; SGSN, serving General Packet Radio Service (GPRS) support node; GGSN, gateway GPRS support node; DSL, digital subscriber line.

Table 5.
Lifetime assumptions of ICT equipment deployed in the radio access network (RAN), core network (CN), and the data and control center (DCC) (based on analytical conclusions and [32, 35]).

delivered to the data concentrator by means of the PLC technology. The forwarding of data from the data concentrator toward the DCC is accomplished by means of cellular mobile communication systems, i.e., GSM and UMTS, for further evaluation and processing purposes. AMI is expected to bring a huge number of advantages with it, like increased reliability and energy efficiency, as well as a thorough insight into the condition of the entire smart grid. This will provide the staff at the DCC with advanced management and monitoring opportunities and enable important remote control functions essential in the course of unusual or unexpected events [34].

The home area networks (HANs) considered here can be seen as an enhancement of the advanced metering infrastructure (AMI). They extend the smart grid idea into the home and enable important home energy management functions. The HAN is used to link various consumer appliances with the home energy management system (HEMS) by means of PLC or, e.g., a low-rate wireless personal area network (LR-WPAN) communication technology like ZigBee. The communication technology deployed will highly depend on the location of the HEMS, i.e., the distance between the HEMS and the various monitored and managed consumer appliances. UDs, i.e., smartphones, tablets, notebooks, DSL modems, and HEMSs, provide users with real-time electricity consumption information by means of a local area network (LAN) and/or wireless local area network (WLAN). This gives them the opportunity to see where and when their electricity consumption is at its peak, providing a way to cut down electricity bills. The link between the HEMS and the utility energy management system (UEMS) is achieved by means of the Internet and the utility web portal [34].

3.2 Description of the scenarios

This section considers various scenarios defined for the environmental sustainability assessment of ICT equipment involved in AMI and HANs. Different assumptions and parameter alterations are defined with the aim to gain more insight into the distribution and development of the ICT equipment exergy consumption. Based on that, meaningful and useful conclusions on the environmental sustainability of the ICT equipment are provided.

The presented scenarios are based on a model developed for the city of Vienna. The embodied exergy consumption (EEC) and operational exergy consumption (OEC) are assessed over a time period from 2020 to 2040, i.e., an operational duration of 20 years is considered. The main assumption is that by the year 2020, the city of Vienna will be equipped with appropriate smart metering, data processing, and forwarding equipment (i.e., smart meters, PLC modems, and data concentrators), required for a correct operation of the AMI application [46]. The equipment needed for an appropriate functioning of the HAN application, e.g., smartphones, tablets, notebooks, digital subscriber line (DSL) modems, and personal computer (PC) towers, is already in existence and widely utilized for various other purposes in almost every (if not every) household. These devices may as well be utilized for a multitude of different home energy management applications in connection with the smart grid concept. The PC tower, for example, may provide the functionality of a home energy management system (HEMS). For that purpose, adequate software programs would be required. Here, however, the HEMS is assumed to be implemented in the form of a server, which is able to support energy management applications of more than one household. The HEMS will most likely be placed with many others near a few houses and/or buildings and will be responsible for the energy management of their various households. This HEMS service could be offered by, e.g., some third-party service provider. It is further assumed that in 2020 only 20% of all households in Vienna will make use of HANs. This percentage is, moreover, assumed to grow to 80% in 2040, as it is expected that not all consumers will agree to deploy HANs in their households. Smartphones, tablets, and notebooks could be equipped with suitable mobile applications and software programs as well, enabling users to visualize their electricity consumption. This would, moreover, lead to an increased energy consumption awareness and enable a greater involvement of consumers in the smart grid concept.

Information on the number of households in Vienna, their expected development, as well as the average number of persons per household is obtained from Statistics Austria. The number of households in Vienna is expected to increase from 927,905 in 2020 to 1,027,846 in 2040 [36]. This corresponds to a yearly average household increase of 4997.05 households. The average number of persons per household during this time period is assumed to be equal to 2. Based on this (and the assumptions regarding the development of HANs between 2020 and 2040 described above), the average yearly increase of households that make use of HANs between 2020 and 2040 is estimated to be approximately equal to 31,834.8. That is, the number of households that deploy HANs will increase from 185,581 in 2020 to 822,277 in 2040. The total traveled distance of extracted and processed raw materials to their manufacturing and assembly location in Shenzhen, Guangdong, in China, is assumed to be equal to 5000 km. From there, the final products are transported over Shanghai, China, and Hamburg, Germany, to the location where they will be deployed, namely, to Vienna, Austria. The total traveled distance of these products was estimated to be equal to 22,403 km. For the end-of-life transportation, a recycling plant in Berlin, Germany, is assumed. The total traveled distance of used up, damaged, and outdated ICT equipment to this location was estimated to be equal to 675 km. The provided distances between the different ICT equipment life cycle stages (i.e., raw material extraction and processing, manufacturing and assembly, operation, recycling and disposal) were estimated by means of the Google Maps route planner. Moreover, various transportation modes (i.e., truck, rail, and ship) between these different locations are considered.

Exergy consumption data of utility equipment (UE), i.e., smart meters, PLC modems, and data concentrators, as well as user devices (UDs), i.e., smartphones, tablets, notebooks (13-inch), and HEMSs, was presented and analyzed in Refs. [37, 38]. The estimation of the exergy consumption of the Apple 15-inch MacBook Pro with Retina Display (w/RD) notebook was based on analytical conclusions as well as the data and information provided in Refs. [39–41]. Exergy consumption data of RAN and CN equipment (including copper and optical fiber cables), as well as that of routers and switches (i.e., rack/edge, aggregation, and core switches), was obtained from Refs. [25, 32, 33]. Network configuration parameters of RAN and CN equipment were based on the data and information provided in Refs. [25, 35]. The evaluation of the exergy consumption of the DSL modem was based on an analytical analysis as well as the information provided in Refs. [42, 43]. Exergy consumption data of the server was based on the data provided in Refs. [8, 13]. The lifetime of BTS and BSC racks is assumed to be equal to 7 and 8 years, respectively, and that of Node B and RNC racks to 8 and 9 years, in that order. The lifetime of both the SGSN and GGSN racks is assumed to be equal to 10 years. The lifetime of DCC routers, CN/DCC switches, and DCC DSL modems is assumed to be equal to 3 years (see also **Table 5**). Data on technology penetration was based on the information obtained from Statistics Austria as well as the forecasts provided in Safaei [32].

3.2.1 Scenario 1: deployment of AMI

This scenario analyzes the environmental sustainability of ICT equipment involved in the advanced metering infrastructure (AMI). The first part of this scenario, i.e., Scenario 1.a, investigates how different utility equipment (UE) lifetime assumptions influence the cumulative embodied exergy consumption (EEC) of the overall model developed for the city of Vienna, over an operational duration of 20 years. The second part of this scenario, i.e., Scenario 1.b, analyzes how the number of smart meters connected to a data concentrator impacts both the cumulative EEC and operational exergy consumption (OEC).

3.2.1.1 Scenario 1a: influence of the utility equipment (UE) lifetime

As the customer and distribution domains of the smart grid will be equipped with a huge number of utility equipment (UE), namely, smart meters, power line communication (PLC) modems, and data concentrators, the means to gain insight into the exergy consumption of this equipment in connection with different lifetime assumptions would prove beneficial. For that purpose, three different use cases (UCs) are defined which assume different lifetimes of the considered UE. The assumptions for these three UCs are listed in **Table 6**.

UC 1 assumes a short lifetime of UE, i.e., smart meters and PLC modems are replaced every 5 years, and the data concentrator even every 3 years. UC 2 and UC 3, on the other hand, assume a longer lifetime of UE. UC 2 defines, moreover, the

UC	UE lifetime (years)		
	Smart meter	**PLC modem**	**Data concentrator**
UC 1: short lifetime	5	5	3
UC 2: medium lifetime	15	10	7
UC 3: long lifetime	20	15	10

PLC, power line communication.

Table 6.
Use case (UC) assumptions for the utility equipment (UE) lifetime.

UC	Number of smart meters per DC
UC 1: 150 smart meters per DC	150
UC 2: 2000 smart meters per DC	2000

Table 7.
Use case (UC) assumptions for different data concentrator (DC) configurations.

basis for Scenario 1.b, which analyzes how the number of smart meters that can be served by a single data concentrator influences the cumulative embodied and operational exergy consumption. It should be noted that the assumed number of smart meters connected to a data concentrator for the present Scenario 1.a equals to 150, which corresponds to the number provided by UC 1 of the following Scenario 1.b (see also UC 1 in **Table 7**). Information on the amount of data traffic per data concentrator, required for the assessment of AMI, was obtained from Luan et al. [44].

3.2.1.2 Scenario 1b: influence of the data concentrator (DC) configuration

The number of smart meters that can be served by a data concentrator can be very high. Up to 2000 smart meters can be linked to a single data concentrator [45]. The present scenario analyzes how the number of smart meters connected to a data concentrator relates to the distribution and development of the cumulative embodied exergy consumption (EEC) and operational exergy consumption (OEC) of the overall system. Two use cases (UCs) with different assumed numbers of smart meters linked to a data concentrator are considered. **Table 7** provides the assumptions for these two UCs. This scenario is based on UC 2 of Scenario 1.a, according to which the lifetime of smart meters, PLC modems, and data concentrators equals to 15, 10, and 7 years, respectively (see also UC 2 in **Table 6**).

UC 1 assumes that 150 smart meters are linked to a data concentrator, whereas UC 2, with 2000 smart meters per data concentrator, represents the upper limit. This scenario will give more insight into the cumulative EEC and OEC in the case of these two UCs. Moreover, UC 1 of the present Scenario 1.b (with 150 smart meters per data concentrator) defines the basis for Scenario 2 described in the following subsection.

3.2.2 Scenario 2: deployment of AMI and HANs

This scenario assesses the environmental sustainability of ICT equipment involved in both the advanced metering infrastructure (AMI) and home area networks (HANs). The first part of this scenario, i.e., Scenario 2.a, analyzes how different lifetimes of utility equipment (UE) as well as user devices (UDs) influence the cumulative embodied exergy consumption (EEC) of the overall system.

Scenario 2.b, on the other hand, investigates how various parameter alterations, e.g., daily charging durations and home energy management usage factors (HEMUFs) of smartphones, tablets, and notebooks, as well as daily uptimes and average loads of home energy management systems (HEMSs), influence the cumulative EEC and operational exergy consumption (OEC) of the overall system. The assumed parameters of these UDs define, thereby, the utilization intensity of HANs. Finally, Scenario 2.c analyzes how the number of households that can be served by a single HEMS, i.e., its configuration, influences the total cumulative exergy consumption of the overall system.

3.2.2.1 Scenario 2.a: influence of the devices' lifetime

The present scenario can be seen as the enhancement of Scenario 1.a, as it considers additionally to utility equipment (UE) also user devices (UDs), i.e., smartphones, tablets, notebooks, home energy management systems (HEMSs), and digital subscriber line (DSL) modems, essential for a proper and easy utilization of home area networks (HANs). As in the case of Scenario 1.a, three use cases (UCs) are defined, which assume different lifetimes of the considered UDs. Each UC of Scenario 2.a is, moreover, related to the respective UC of Scenario 1.a. This can be seen in **Table 8**, which provides the assumed lifetimes for both the UDs and UE for the three considered UCs.

The first UC assumes a short lifetime of UDs as well as UE. UC 2 defines the most probable case considering the lifetimes of UDs and UE. UC 3 assumes an extended lifetime for both the UDs and UE, i.e., their replacement period is longer than that of the first two UCs. Based on these UCs, more information on the exergy consumption distribution and its development will be provided. UC 2 defines, moreover, the basis for Scenario 2.b as well as Scenario 2.c, which assess how different UDs' parameters, e.g., daily charging durations and home energy management usage factors (HEMUFs) of smartphones, tablets, and notebooks, daily uptimes and average loads of HEMSs, as well as HEMS configurations, influence the cumulative embodied and operational exergy consumption. It should be noted that the assumed daily charging durations and HEMUFs of smartphones, tablets, and notebooks, daily uptimes and average loads of HEMSs, as well as HEMS configurations for the present Scenario 2.a correspond to those of UC 2 of Scenario 2.b as well as Scenario 2.c discussed in the following two parts of this subsection.

UC	Lifetime of the user devices (UDs) (years)				
	Smartphone	Tablet	Notebook	HEMS	DSL modem
UC 1: short lifetime	1	1	2	2	3
UC 2: medium lifetime	2	2	3	4	5
UC 3: long lifetime	4	4	6	6	10
UC	**Lifetime of utility equipment (UE) (years)**				
	Smart meter	**PLC modem**	**Data concentrator**		
UC 1: short lifetime	5	5	3		
UC 2: medium lifetime	15	10	7		
UC 3: long lifetime	20	15	10		

HEMS, home energy management system; DSL, digital subscriber line; PLC, power line communication.

Table 8.
Use case (UC) assumptions for the user devices (UDs) and utility equipment (UE) lifetime.

3.2.2.2 Scenario 2.b: influence of the home area network (HAN) utilization

This scenario analyzes how different daily charging durations and home energy management usage factors (HEMUFs) of smartphones, tablets, and notebooks, as well as daily uptimes and average loads of home energy management systems (HEMSs), influence the exergy consumption of the overall system. For that purpose, three use cases (UCs) are defined which assume different utilization intensities of home area networks (HANs). The assumed daily charging durations and HEMUFs of smartphones, tablets, and notebooks, as well as the daily uptimes and average loads of HEMSs for these three UCs, are provided in **Tables 9–11**. This scenario is based on UC 2 of Scenario 2.a, according to which the lifetime of smartphones tablets, notebooks, HEMSs, and DSL modems equals to 2, 2, 3, 4, and 5 years, respectively, and that of smart meters, PLC modems, and data concentrators to 15, 10, and 7 years.

UC 1 corresponds to a high utilization of HANs, which is associated with longer smartphone, tablet, and notebook daily charging durations and higher HEMUFs, as well as a higher average HEMS load. UC 3, on the other hand, is associated with a low utilization of HANs. UC 2 represents the most probable usage pattern of HANs. Moreover, UC 2 of the present scenario defines the basis for Scenario 2.c described in the following part of this subsection. The daily uptime of HEMSs and DSL modems is set to 100% (i.e., 24 h) for all three UCs, however, with varying average loads (see **Table 11**). Further, the average load of the DSL modem is assumed to be proportional to that of the HEMS for all the three considered UCs. It is important to

UC	Daily charging duration (h)		
	Smartphone	Tablet	Notebook
UC 1: high utilization	4	4	6
UC 2: medium utilization	2	2	3
UC 3: low utilization	1	1	2

Table 9.
Use case (UC) assumptions for daily charging duration.

UC	HEMUF (h)			Daily uptime (%)	
	Smartphone	Tablet	Notebook	HEMS	DSL modem
UC 1: high utilization	4	4	4	100	100
UC 2: medium utilization	2	2	2	100	100
UC 3: low utilization	1	1	1	100	100

Table 10.
Use case (UC) assumptions for the home energy management usage factor (HEMUF) as well as HEMS and DSL modem daily uptime.

UC	Average HEMS load (%)
UC 1: high utilization	80
UC 2: medium utilization	50
UC 3: low utilization	20

Table 11.
Use case (UC) assumptions for the average home energy management system (HEMS) load.

UC	Number of households per HEMS
UC 1: 10 households per HEMS	10
UC 2: 20 households per HEMS	20
UC 3: 100 households per HEMS	100

Table 12.
Use case (UC) assumptions for different home energy management system (HEMS) configurations.

mention that the assumed number of households that can be served by a single HEMS for the present Scenario 2.b equals to 20 households, which corresponds to the number provided by UC 2 of the following Scenario 2.c (see also UC 2 in **Table 12**).

3.2.2.3 Scenario 2.c: influence of the configuration of the home energy management system (HEMS)

The last scenario analyzes how the number of households that can be served by a single home energy management system (HEMS) influences the cumulative embodied exergy consumption (EEC) and operational exergy consumption (OEC) of the overall system. As already mentioned several times across this chapter, the HEMS is assumed to be implemented in the form of a 2-unit (2 U) rack-mounted server, which is placed at a convenient location near the various homes and/or buildings it is responsible for to manage. Moreover, the HEMS is considered to be equipped with all the necessary software programs required for a correct and reliable functioning of home energy management applications, i.e., home area networks (HANs). Three different use cases (UCs) are considered which assume different HEMS configurations, i.e., numbers of households it can serve. The assumptions for this three UCs are provided in **Table 12**. This scenario is based on UC 2 of Scenario 2.a as well as Scenario 2.b, which assume a medium lifetime of user devices (UDs) and utility equipment (UE), as well as a medium utilization intensity of HANs.

UC 1 and UC 2 assume that the HEMS can support home energy management applications of 10 and 20 households, respectively. UC 3, on the other hand, assumes that 100 households can be served by a single HEMS. Based on this scenario, more information on the cumulative EEC and OEC for the three defined UCs will be provided. This provides, moreover, means to indicate the most exergy consumption-related category in the case of these three UCs, i.e., whether the EEC or OEC dominates over the considered operational duration of 20 years.

4. Major findings of the E-LCA study

It can be argued that for all of the considered scenarios, the customer and distribution domains are the most exergy-consuming domains. For that reason, they have the highest potential to negatively impact the environment, i.e., they are closely associated with environmental sustainability issues.

Considering only the advanced metering infrastructure (AMI) scenarios (i.e., Scenarios 1.a and 1.b), the utility equipment (UE) was ascertained to be the ICT equipment category group related to the highest cumulative embodied exergy consumption (EEC) as well as operational exergy consumption (OEC). The contribution of the data and control center (DCC) equipment to the total cumulative OEC was determined to be relatively high (i.e., about 20%), taking into account that the number of DCC equipment is much less than that of UE. The cumulative EEC of the DCC equipment was ascertained to be around 1% and is for that reason almost negligible. The contribution of radio access network (RAN) and core network (CN) equipment to the total cumulative EEC and OEC turned out to be very low compared to that of the UE and DCC equipment (i.e., lower than 1%). Such a low contribution of RAN and CN equipment to the total cumulative exergy consumption of the overall system arises from the fact that it is used relatively shortly for AMI throughout the day, i.e., for only a few minutes daily.

Figures 9 and **10** depict the share of the embodied exergy consumption (EEC) and operational exergy consumption (OEC) to the total cumulative exergy consumption of the overall system for the Scenarios 1.a and 1.b, respectively, at the end

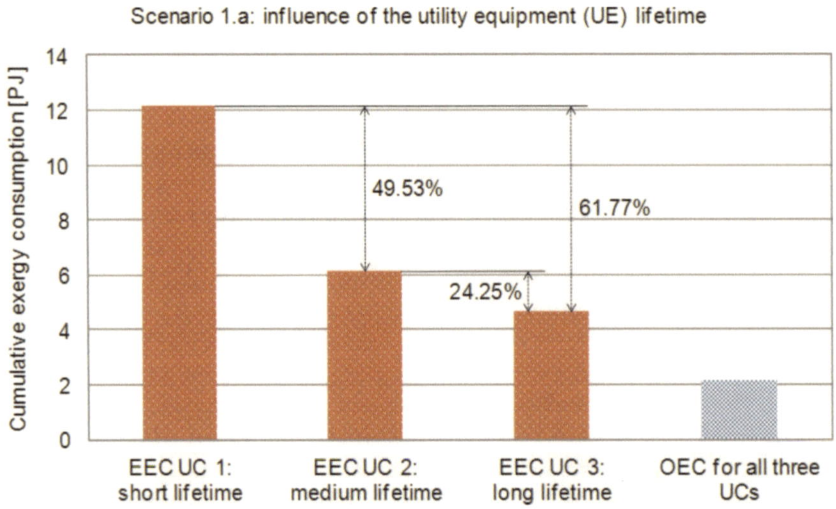

Figure 9.
Distribution of the cumulative embodied exergy consumption (EEC) and operational exergy consumption (OEC) of Scenario 1.a for the three defined use cases (UCs) with different utility equipment (UE) lifetimes, after an operational duration of 20 years.

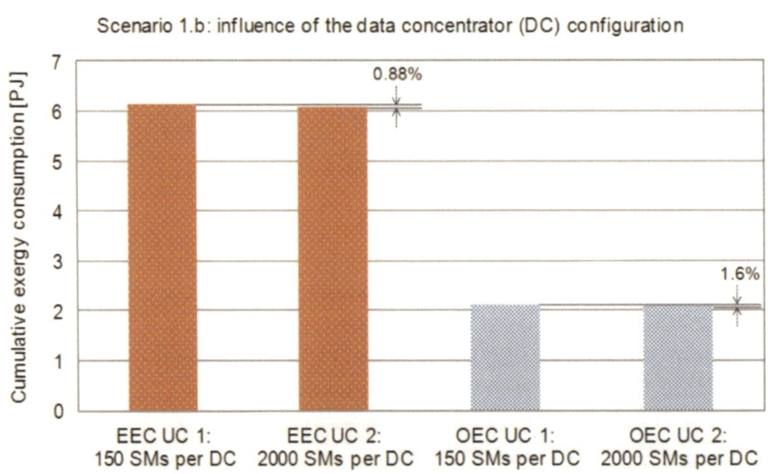

Figure 10.
Distribution of the cumulative embodied exergy consumption (EEC) and operational exergy consumption (OEC) of Scenario 1.b for the two defined use cases (UCs) with different numbers of smart meters (SMs) per data concentrator (DC), after an operational duration of 20 years.

of the operational duration of 20 years. As can be seen from **Figure 9**, the cumulative EEC differs significantly for different utility equipment (UE) lifetimes. The cumulative OEC is the same for all the three use cases (UCs) and is, moreover, lower than their cumulative EEC. A longer UE lifetime decreases considerably the cumulative EEC (see the provided percentages in **Figure 9**). A reduction of up to about 61.77% at the end of the operational duration of 20 years is possible, if the lifetime of smart meters, power line communication (PLC) modems, and data concentrators is extended to 20, 15, and 10 years (i.e., UC 3), respectively, in contrast to 5, 5, and 3 years (i.e., UC 1). Extending the lifetime of smart meters, PLC modems, and data concentrators to 15, 10, and 7 years (i.e., UC 2), respectively, reduces the cumulative EEC by about 49.53% compared to UC 1. Moreover, the difference between the cumulative EEC of UC 2 and UC 3 at the end of the operational duration of 20 years corresponds to approximately 24.25%.

From **Figure 10** it is evident that the number of smart meters (SMs) that can be served by a single data concentrator (DC) does not have a strong impact on the cumulative EEC as well as OEC. The difference between the cumulative EEC of UC 1 (which assumes that 150 smart meters are linked to a data concentrator) and UC 2 (which assumes 2000 smart meters per data concentrator) corresponds to approximately 0.88%, which is not that high. The difference between the cumulative OEC of these two UCs is not that large as well and equals to about 1.6%. It can be concluded that the number of smart meters connected to a data concentrator does not have a large influence on the cumulative EEC and OEC of the overall system. The impact of the UE lifetime has a much stronger impact on the total cumulative exergy consumption of the overall system (see the percentages provided in **Figure 9**).

In the case of the advanced metering infrastructure (AMI) and home area network (HAN) scenarios (i.e., Scenarios 2.a–2.c), the utility equipment (UE) was determined to be the ICT equipment category group with the highest share to the cumulative embodied exergy consumption (EEC). The user devices (UDs) were ascertained to be the next largest contributor to the overall cumulative EEC. The assessment of the cumulative operational exergy consumption (OEC), however, revealed a large dependence on the utilization intensity of HANs. That is, for a high and medium utilization intensity of HANs, it was shown that the UE represents the most dominant ICT equipment category group until a certain time point, from where on the UDs become the ICT equipment category group with the highest exergy expenditure and with that the category group closely linked to environmental sustainability issues. For a low utilization intensity of HANs, however, the UE turned out to be the ICT equipment category group with the highest contribution to the cumulative OEC over the entire operational duration of 20 years. In this case, it is the UE that is closely associated with increased environmental sustainability issues. The share of radio access network (RAN) and core network (CN) equipment to the total cumulative exergy consumption of the overall system was determined to be less dominant for all the three considered scenarios, as this equipment is utilized for AMI for only a few minutes daily. Therefore, it can be argued that the RAN and CN equipment is associated with lower environmental sustainability issues. The contribution of the data and control center (DCC) equipment to the total cumulative exergy consumption of the overall system turned out to be larger than that of the RAN and CN equipment, but lower than that of the UE and UDs.

Figures 11–13 provide the distribution of the cumulative embodied exergy consumption (EEC) and operational exergy consumption (OEC) to the total cumulative exergy consumption of the overall system for Scenarios 2.a–2.c, respectively, after the operational duration of 20 years.

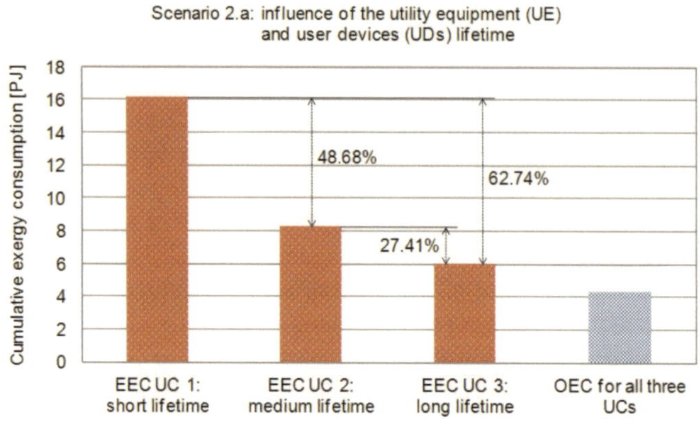

Figure 11.
Distribution of the cumulative embodied exergy consumption (EEC) and operational exergy consumption (OEC) of Scenario 2.a for the three defined use cases (UCs) with different utility equipment (UE) and user device (UD) lifetimes, after an operational duration of 20 years.

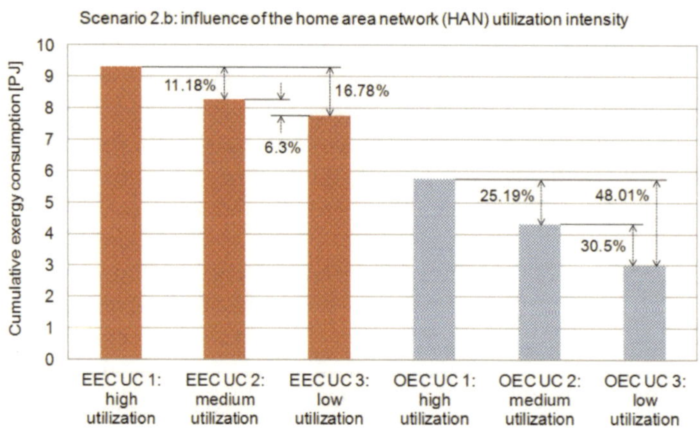

Figure 12.
Distribution of the cumulative embodied exergy consumption (EEC) and operational exergy consumption (OEC) of Scenario 2.b for the three defined use cases (UCs) with different home area network (HAN) utilization intensities, after an operational duration of 20 years.

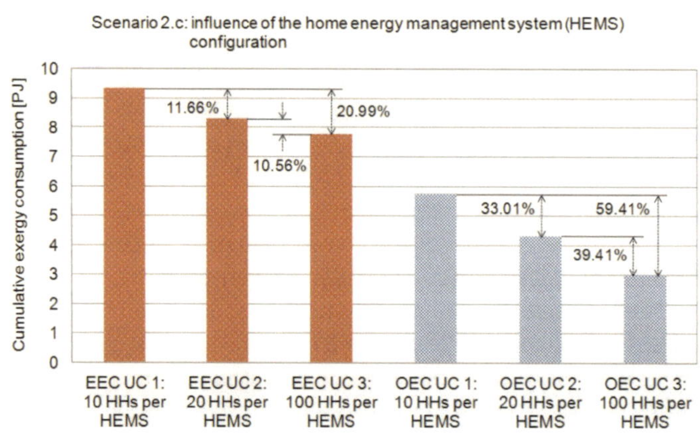

Figure 13.
Distribution of the cumulative embodied exergy consumption (EEC) and operational exergy consumption (OEC) of Scenario 2.c for the three defined use cases (UCs) with different numbers of households (HHs) per home energy management system (HEMS), after an operational duration of 20 years.

From **Figure 11** it is clear that an increased lifetime of user devices (UDs) and utility equipment (UE) leads to a significant reduction of the cumulative EEC of the overall system, at the end of the operational duration of 20 years. Moreover, the cumulative OEC is the same for all three use cases (UCs) and turns out to be lower than the cumulative EEC for all the three considered UCs. A decrease of about 48.68% can be achieved if the short lifetime of UDs and UE (i.e., UC 1) is extended to a medium lifetime (i.e., UC 2). Extending the lifetime of UDs and UE further (i.e., UC 3), an even larger cumulative EEC reduction of about 62.74% becomes possible. Furthermore, the cumulative EEC difference between UC 2 and UC 3 corresponds to approximately 27.41%.

As can be seen from **Figure 12**, an increased utilization intensity of home area networks (HANs) leads to a significant increase of the cumulative EEC but even more notably the cumulative OEC. Nevertheless, the cumulative OEC is lower than the cumulative EEC in the case of all the three considered use cases (UCs). The cumulative EEC difference between the high and medium utilization intensities of HANs (i.e., UC 1 and UC 2) equals to about 11.18%. Comparing UC 1 and UC 3 (i.e., low utilization of HANs) provides a cumulative EEC difference of approximately 16.78%. Moreover, the cumulative EEC difference between UC 2 and UC 3 equals to about 6.3%. From **Figure 12** it is also obvious that the usage pattern of HANs has a much higher impact on the cumulative OEC than on the cumulative EEC. The difference between the high and medium utilization of HANs (i.e., UC 1 and UC 2) shows an approximately 25.19% difference of the cumulative OEC, at the end of the operational duration of 20 years. The difference of the cumulative OEC between UC 1 and UC 3 (which assumes a low utilization of HANs) is even more pronounced and corresponds to about 48.01%. Furthermore, a comparison of the cumulative OEC between UC 2 and UC 3 exhibits a difference of approximately 30.5%. Such a large cumulative OEC difference between the three considered UCs is mainly associated with the different average home energy management system (HEMS) loads assumed for each of these UCs (see also **Table 11**). That is, an increased HAN usage pattern is related to higher average HEMS loads and leads therefore to an increase of the cumulative OEC. The contributions of the daily charging durations as well as the respective home energy management usage factors (HEMUFs) of smartphones, tablets, and notebooks have a minor impact on the total cumulative OEC of the overall system compared to that of the average HEMS load.

From **Figure 13** it is obvious that the number of households (HHs) that can be served by a single home energy management system (HEMS), i.e., its configuration, has a large impact on the cumulative EEC as well as OEC of the overall system. Moreover, the cumulative OEC is ascertained to be lower than the cumulative EEC for all the three considered use cases (UCs). A cumulative EEC reduction of about 11.66% can be achieved if 20 households are linked to a single HEMS (i.e., UC 2) instead of 10 households (i.e., UC 1). Increasing the number of households from 10 (i.e., UC 1) to 100 (i.e., UC 3) leads to an even larger cumulative EEC reduction of approximately 20.99%. Furthermore, the difference between the cumulative EEC of UC 2 and UC 3 equals to about 10.56%. Taking a look at **Figure 13**, it can be seen that the HEMS configuration has also a considerable, and an even larger, impact on the cumulative OEC of the overall system. The difference between the cumulative OEC of UC 1 and UC 2 (i.e., for 10 and 20 households per HEMS, respectively) amounts approximately to 33.01%. If the number of households that can be managed by a single HEMS is increased from 10 to 100, the cumulative OEC difference becomes even larger and amounts about 59.41%. Moreover, a cumulative OEC reduction of approximately 39.41% can be attained if 100 households are managed by a single HEMS instead of 20.

5. Conclusions

The exergy-based life cycle assessment (E-LCA) of utility equipment (UE), i.e., smart meters, power line communication (PLC) modems, and data concentrators, revealed that the lifetime embodied exergy consumption (EEC) represents the most exergy consumption-related category. Based on that, it can be argued that the lifetime EEC is the exergy consumption category closely associated with environmental sustainability issues. The lifetime operational exergy consumption (OEC) was ascertained to have a much lower impact on the environment. Furthermore, the manufacturing and assembly stage turned out to be the most dominant life cycle stage in the case of the considered UE. The processor was additionally ascertained to be the most exergy consumption-related UE component.

Similar to the results for the utility equipment, the analysis has shown that also for user devices such as smartphones, tablets, and notebooks the most exergy-consumption related category is the embodied exergy consumption (EEC). In particular, the manufacturing and assembly stages turned out to be the most dominant life cycle stage in the case of these user devices (UDs). The processor was also here ascertained to be the most exergy consumption-related component. Just as in the case of the UE, the lifetime operational exergy consumption (OEC) of the smartphone, tablet, and notebook was determined to be the exergy consumption category associated with lower (i.e., almost negligible) environmental sustainability issues, when compared to the lifetime EEC of these devices. However, the E-LCA of the home energy management system (HEMS) revealed that the lifetime OEC represents the most exergy consumption-related category, i.e., the category closely associated with environmental sustainability issues.

The exergy-based life cycle assessment (E-LCA) of the overall model developed for the city of Vienna revealed that the customer and distribution domains are the most exergy consumption-related domains, i.e., these domains are closely linked to environmental sustainability issues. Scenarios considering only the advanced metering infrastructure (AMI) ascertained the utility equipment (UE) as the ICT equipment category group leading to the highest cumulative embodied exergy consumption (EEC) as well as operational exergy consumption (OEC). The share of the data and control center (DCC) equipment to the total cumulative OEC was determined to be relatively high (i.e., about 20%), considering the much lower number of DCC equipment compared to that of the UE. The share of radio access network (RAN) and core network (CN) equipment to the total cumulative exergy consumption of the overall system was ascertained to have a much lower impact on the environment than the UE and DCC equipment. The reason for such a low contribution of RAN and CN equipment to the overall cumulative exergy consumption arises from the fact that it is used relatively shortly for the AMI application throughout the day. Moreover, the cumulative EEC was ascertained to be the most exergy consumption-related category over the entire operational duration of 20 years. For that reason, it is associated with increased environmental sustainability issues. It was shown that an increase of the UE lifetime has a strong impact on the cumulative EEC. Increasing the lifetime of smart meters, power line communication (PLC) modems, and data concentrators from 5, 5, and 3 years, respectively, to 15, 10, and 7 years, in that order, results at the end of the operational duration of 20 years in a cumulative EEC reduction of about 49.53%. A cumulative EEC decrease of approximately 61.77% is possible if the smart meter, PLC modem, and data concentrator are replaced every 20, 15, and 10 years, respectively, instead of every 5, 5, and 3 years. Moreover, it was shown that the number of smart meters that can be served by a single data concentrator does not have a strong impact on the total cumulative exergy consumption of the overall system. A reduction of the cumulative EEC of approximately 0.88% after the operational duration of 20 years can be attained if 2000 smart meters are linked to a data concentrator instead of 150. The reduction of the cumulative OEC for these two different data concentrator configurations is not that large as well and corresponds to about 1.6%.

The exergy-based life cycle assessment (E-LCA) of the advanced metering infrastructure (AMI) and home area network (HAN) scenarios ascertained the utility equipment (UE) as the ICT equipment category group with the largest contribution to the cumulative embodied exergy consumption (EEC). The user devices (UDs) were determined to be the next largest contributor to the total cumulative EEC of the overall system. However, the analysis and evaluation of the cumulative operational exergy consumption (OEC) showed that the HAN utilization intensity determines which ICT equipment category group is the most dominant one. It turned out that for a high and medium HAN utilization intensity, the UE contributes the most to the cumulative OEC until a specific time point, from where on the UDs become the most dominant ICT equipment category group, i.e., the category group closely associated with increased environmental sustainability issues. For a low utilization intensity of HANs, the UE was ascertained to be the ICT equipment category group with the highest contribution to the cumulative OEC over the entire operational duration of 20 years, i.e., the category group closely related to environmental sustainability issues. The contribution of radio access network (RAN) and core network (CN) equipment to the overall cumulative exergy consumption was determined to be less dominant. This result is based on the fact that this equipment is utilized for AMI for merely a few minutes daily. For that reason, this equipment is associated with lower environmental sustainability issues. The share of the data and control center (DCC) equipment to the total exergy consumption of the overall system was ascertained to be larger than that of the RAN and CN equipment, but lower when compared to that of the UDs and UE. Moreover, the contribution of the DCC equipment to the cumulative OEC is more pronounced than its share to the cumulative EEC. As in the case of the AMI scenarios, an increase of the UDs' as well as UE lifetime leads to a considerable decrease of the cumulative EEC. Further, the cumulative EEC turned out to be the most exergy consumption-related category over the entire operational duration of 20 years, i.e., the category group closely linked to environmental sustainability issues, just as in the case of the scenarios considering only AMI. It was determined that an increase of the lifetime of UDs and UE results in a significant reduction of the cumulative EEC. A decrease of the cumulative EEC of about 48.68% at the end of the operational duration of 20 years is possible, if the UDs' and UE lifetime is extended from a short lifetime to a medium lifetime. A further extension of the UDs' and UE lifetime resulted in an even larger cumulative EEC reduction of approximately 62.74%. The assessment of the utilization intensity of home energy management applications, i.e., HANs, revealed a large influence on the cumulative EEC and, even more, the cumulative OEC. It was shown that an increase from a low utilization intensity of HANs (i.e., average HEMS load equals to 20%) to a medium utilization intensity of HANs (i.e., average HEMS load equals to 50%) leads to an approximately 11.18% cumulative EEC increase as well as an approximately 25.19% cumulative OEC increase at the end of the operational duration of 20 years. A further increase of the utilization intensity to a high utilization of HANs (i.e., the average HEMS load corresponds to 80%) showed an even larger cumulative EEC and OEC increase of about 16.78% and 48.01%, respectively, compared to a low utilization intensity of HANs. Furthermore, it was ascertained that the number of households that can be managed by a single home energy management system (HEMS), i.e., its configuration, has a considerable impact on the cumulative EEC as well as OEC of the overall system. That is, an increase of households that can be served by a single HEMS from 10 to 20 households showed, after the operational duration of 20 years, a cumulative EEC and OEC decrease of about 11.66 and 33.01%, respectively. A further increase of households linked to a single HEMS from 10 to 100 revealed an even larger cumulative EEC and OEC reduction of approximately 20.99% and 59.41%, respectively.

Finally, it can be concluded that the lifetime of utility equipment (UE) as well as user devices (UDs) has a strong influence on the cumulative embodied exergy consumption (EEC). An increase of the UE's and UDs' lifetime results in a considerable decrease of the cumulative EEC. Moreover, the utilization intensity of home energy management applications, i.e., home area networks (HANs), defined especially by the average home energy management system (HEMS) load, revealed a large impact on the cumulative operational exergy consumption (OEC). The daily charging durations as well as home energy management usage factors (HEMUFs) of smartphones, tablets, and notebooks have a minor influence on the cumulative OEC when compared to the average HEMS load. Nevertheless, the cumulative EEC turned out to be related to the largest exergy consumption for all of the studied scenarios and over the entire considered operational duration of 20 years. For that reason, it can be concluded that the most dominant life cycle stages of the considered overall system relate to those involved in the raw material extraction and processing, manufacturing and assembly, recycling and disposal, as well as the transportation. The operation phase of the overall system turned out to be the less dominant one. Therefore, it can be concluded that the EEC is closely associated with environmental effects, i.e., increased environmental sustainability issues.

Author details

Vedad Mujan[1] and Slavisa Aleksic[2*]

1 Vienna University of Technology, Vienna, Austria

2 Leipzig University of Telecommunications (HfTL), Leipzig, Germany

*Address all correspondence to: aleksic@hft-leipzig.de

References

[1] U.S. Energy Information Administration (EIA). International Energy Outlook 2013. DOE/EIA-0484 (2013). 2013. Available from: http://www.eia.gov/forecasts/ieo/pdf/0484(2013).pdf [Accessed: 11 March 2015]

[2] International Energy Agency (IEA). World Energy Outlook 2013. London. 2012. Available from: http://www.worldenergyoutlook.org/pressmedia/recentpresentations/londonnovember12.pdf [Accessed: 11 March 2015]

[3] International Telecommunication Union (ITU). Boosting energy efficiency through Smart Grids. 2012. Available from: http://www.itu.int/dms_pub/itu-t/oth/4B/01/T4B010000050001PDFE.pdf [Accessed: 11 March 2015]

[4] OECD. ICT Applications for the Smart Grid: Opportunities and Policy Implications. OECD Digital Economy Papers, No. 190. OECD Publishing; 2012. DOI: 10.1787/5k9h2q8v9bln-en [Accessed: 11 March 2015]

[5] European Commission. Information Society, ICT for Sustainable Growth, "Smart Grids". 2009. Available from: http://ec.europa.eu/information_society/activities/sustainable_growth/grids/index_en.htm [Accessed: 26 March 2015]

[6] Aleksic S. Energy, entropy and exergy in communication networks. Entropy. 2013;**15**(10):4484-4503

[7] Lettieri DJ, Hannemann CR, Carey VP, Shah AJ. Lifetime exergy consumption as a sustainability metric for information technologies. In: Proceedings of the 2009 IEEE International Symposium on Sustainable Systems and Technology. IEEE Computer Society. 2009. pp. 1-6

[8] Hannemann CR, Carey VP, Shah AJ, Patel C. Lifetime exergy consumption as a sustainability metric for enterprise servers. In: Proceedings of ASME 2nd International Conference on Energy Sustainability. Jacksonville, Florida USA; 2008. pp. 1-8

[9] Global e-Sustainability Initiative (GeSI). ICT Sustainability through Innovation. GeSI Activity Report. 2009. Available from: http://gesi.org/files/Reports/ICT%20Sustainability%20through%20Innovation-%20GeSI%20Activity%20Report%202009.pdf [Accessed: 26 March 2015]

[10] Rosen MA, Dincer I, Kanoglu M. Role of exergy in increasing efficiency and sustainability and reducing environmental impact. Energy Policy. 2008;**36**(1):128-137

[11] UN Documents. Gathering a body of global agreements. In: Report of the World Commission on Environment and Development: Our Common Future. 1987. Available from: http://www.un-documents.net/wced-ocf.htm [Accessed: 26 March 2015]

[12] Drexhage J, Murphy D, International Institute for Sustainable Development (IISD). Background Paper. Sustainable Development: From Brundtland to Rio 2012. United Nations Headquarters, New York; 2010. Available from: http://www.un.org/wcm/webdav/site/climatechange/shared/gsp/docs/GSP1-6_Background%20on%20Sustainable%20Devt.pdf [Accessed: 26 March 2015]

[13] David J. Lettieri Expeditious data center sustainability, flow, and temperature modeling: Life-cycle exergy consumption combined with a potential flow based, Rankine vortex superposed, predictive method [doctoral thesis]. University of California, Berkeley; 2012

[14] ISO 14040:2006. Environmental Management—Life Cycle Assessment—Principles and Framework. 2007.

Available from: http://www.iso.org/iso/catalogue_detail%3Fcsnumber%3D37456 [Accessed: 27 March 2015]

[15] ISO 14040, International Standard. Environmental Management—Life Cycle Assessment—Principles and Framework. 1st edn. 1997. Available from: http://web.stanford.edu/class/cee214/Readings/ISOLCA.pdf [Accessed: 27 March 2015]

[16] SimaPro. World's Leading LCA Software. 2013. Available from: http://www.pre-sustainability.com/simapro [Accessed: 27 March 2015]

[17] Yale Center for Environmental Law and Policy, Yale University, Center for International Earth Science Information Network, Columbia University. 2005 Environmental Sustainability Index. Benchmarking National Environmental Stewardship. 2005. Available from: http://www.yale.edu/esi/ESI2005_Main_Report.pdf [Accessed: 26 March 2015]

[18] Frangopoulos CA, editor. Exergy, Energy System Analysis and Optimization, Exergy and Thermodynamic Analysis. Vol. I. Oxford, UK: EOLSS Publications; 2009. p. 298

[19] Herms S. Exergy flows in product life cycles, analyzing thermodynamic improvement potential of cardboard life cycles [master thesis]. Delft, Netherlands: Delft University of Technology, Leiden University; 2011

[20] Terzi R. Application of exergy analysis to energy systems. In: Taner T, editor. Application of Exergy. London, UK: IntechOpen; 2018. pp. 109-123

[21] Thinkstep GaBi. 2014. Available from: http://www.gabi-software.com/solutions/life-cycle-assessment/ [Accessed: 27 March 2015]

[22] EarthShift. Building pathways to sustainability. EarthSmart. 2014. Available from: http://www.earthshift.com/EarthSmart [Accessed: 27 March 2015]

[23] Manrique Delgado B, Cao S, Hasan A, Sirén K. Energy and exergy analysis of prosumers in hybrid energy grids. Building Research & Information. 2017;**46**(6):668-685

[24] Salehi N, Mahmoudi M, Bazargan A, McKay G. Exergy and life cycle-based analysis. In: Hussain CM, editor. Handbook of Environmental Materials Management. Cham: Springer; 2018

[25] Aleksic S, Safaei M. Exergy consumption of cloud computing: A case study. Invited Paper. In: NOC 2014. Milan, Italy; 2014. pp. 1-6

[26] Malmodin J, Lundén D, Moberg Å, Andersson G, Nilsson M. Life cycle assessment of ICT. Journal of Industrial Ecology. 2014;**18**(6):829-845

[27] Moghaddam RF, Moghaddam FF, Dandres T, Lemieux Y, Samson R, Cheriet M. Challenges and complexities in application of LCA approaches in the case of ICT for a sustainable future. In: ICT for Sustainability 2014 (ICT4S 2014), Stockholm, Sweden. 2014. pp. 155-164

[28] Hischier R, Coroama VC, Schien D, Achachlouei MA. Grey energy and environmental impacts of ICT hardware. In: Hilty LM, Aebischer B, editors. ICT Innovations for Sustainability, Advances in Intelligent Systems and Computing. Vol. 310. Cham: Springer; 2014. pp. 171-189

[29] Andrae ASG, Vaija MS. To which degree does sector specific standardization make life cycle assessments comparable?—The case of global warming potential of smartphones. Challenges. 2014;**5**:409-429

[30] iPhone 4S. Environmental Report. 2011. Available from: http://www.apple.com/environment/reports/docs/iPhone

4S_Product_Environmental_Report_2011.pdf [Accessed: 28 March 2015]

[31] iPhone 5C. Environmental Report. 2013. Available from: https://www.apple.com/la/environment/reports/docs/iPhone5c_product_environmental_report_sept2013.pdf [Accessed: 28 March 2015]

[32] Safaei M. Exergy-based life cycle assessment (E-LCA) of cloud computing [master thesis]. Vienna, Austria: Vienna University of Technology, Institute of Telecommunications; 2013

[33] Aleksic S, Safaei M. Exergy based analysis of radio access networks. Invited Paper. In: Proceedings of ICEAA—IEEE APWC—EMS 2013. 2013. pp. 1-8

[34] Budka KC, Deshpande JG, Thottan M. Communication Networks for Smart Grids: Making Smart Grid Real. Computer Communications and Networks, Springer; 2014

[35] Scharnhorst W. Life cycle assessment of mobile telephone networks, with focus on the end-of-life phase [dissertation]. Lausanne: EPFL; 2006

[36] Statistics Austria. Private households forecastes, Privathaushalte 2011–2060 nach Bundesländern. 2014. Available from: http://www.statistik.at/web_de/statistiken/bevoelkerung/demographische_prognosen/haushalts_und_familienprognosen/023529.html [Accessed: 29 March 2015]

[37] Aleksic S, Mujan V. Exergy cost of information and communication equipment for smart metering and smart grids. Sustainable Energy, Grids and Networks. 2018;**14**:1-11

[38] Aleksic S, Mujan V. Life cycle based analysis of ICT equipment for advanced metering infrastructure. In: 13th International Conference on Telecommunications (ConTEL). Graz; 2015. pp. 1-7

[39] MacBook Pro. 15-inch with Retina Display. 2014. Available from: http://www.apple.com/macbook-pro/specs-retina/ [Accessed: 29 March 2015]

[40] Apple Inc. 15-inch MacBook Pro with Retina Display. Environmental Report. 2014. Available from: https://www.apple.com/environment/reports/docs/15inch_MacBookPro_wRetina_PER_July2014.pdf [Accessed: 29 March 2015]

[41] iFixit. MacBook Pro 15" Retina Display Late 2013 Teardown. 2013. Available from: https://www.ifixit.com/Teardown/MacBook+Pro+15-%20Inch+Retina+Display+Late+2013+Teardown/18696 [Accessed: 29 March 2015]

[42] DM602 ADSL Modem Router, Netgear. 2004. Available from: http://kbserver.netgear.com/datasheets/DM602_Datasheet%28noZKS%29_23Feb2004.pdf [Accessed: 29 March 2015]

[43] Inside Gadgets. View the electronics that's inside your gadgets. In: Inside the Netgear DM602 ADSL Modem. 2010. Available from: https://insidegadgets.wordpress.com/2010/10/23/inside-the-netgear-dm602-adsl-modem/#more-551 [Accessed: 29 March 2015]

[44] Luan W, Sharp D, LaRoy S. Data traffic analysis of utility smart metering network. In: Power and Energy Society General Meeting (PES). Vancouver, BC, Canada: IEEE; 2013. pp. 1-4

[45] Punya Prakash. Texas Instruments, White Paper. Data concentrators: The core of energy and data management. Available from: http://www.ti.com/lit/wp/spry248/spry248.pdf [Accessed: 29 March 2015]

[46] Hauth E, Hierzinger R, Seisser O. Smart Services – Smarte Dienstleistungen in Smarten Märkten. In: Final Report e!Mission.at, FFG; 2014. pp. 1-40

Section 3

Use of Advanced Communication and Control in Smartgrids

Chapter 7

Communications for Exploiting Flexible Resources in the Framework of Smart Grids in Islands

Javier Rodríguez-García, David Ribó-Pérez, Carlos Álvarez-Bel and Manuel Alcázar-Ortega

Abstract

Although being among the least responsible for climate change, islands are in great threat due to it. The decarbonisation of the power system arises as a key factor to ensure adaptation and mitigation to it. Islands' characteristics make renewable electrification a challenge. Most islands are isolated systems with low levels of inertia that require stability for ensuring security of supply. Therefore, the potential of smart grids and flexible resources must be fully exploited to ensure a viable integration of renewable energy sources. In this vein, it is necessary to evolve the system including demand response, batteries and electric transport to increase the share of renewables. However, all these elements require a reliable communication architecture to be deployed. A communication architecture is hereby presented and applied to Galapagos for exploiting flexible resources. Different protocols have been selected to interoperate flexible resources integrated on the system. Each of them tries for each application to standardise and ensure the largest functionalities available. The deployment of smart grids in islands can reduce their carbon footprint as it is validated with a case study in Santa Cruz, Galapagos. This system proves to ensure the energy balance in a viable way, in technical, economic and environmental terms.

Keywords: smart grids, renewable generation, communication systems, flexible resources, islands, Galapagos

1. Introduction

Electric power business used to be a very traditional sector with a very well-established structure both in the physical and market layers. Depletion of fossil fuels, climate change, the boom of renewables and communications has forced a change in the ways in which the electricity is generated, consumed and traded. Traditional power systems were formed by large centralised generators, a very large and extended transmission grid to connect the generation and the load sites. Generation arrived to costumers through a final distribution grid, which was usually operated in radial form to feed the large customers and small customers in low voltage (LV) [1]. However, the new time in the power sector are characterised by a large increment in the renewable generation. The main technologies are solar photovoltaic (PV) and wind generation. This generation can be concentrated in medium/large capacity plants or be very distributed. The so-called distributed generation (DG) is much smaller in size (commonly some tens to hundreds of kilowatts) and is usually connected to LV or medium voltage (MV). This generation can be owned by companies or individual customers that may also install some small generation. Moreover, the electrification of transport and the rise of information and communication technology are increasing the possibility to take advantage of customer's flexibility in the consumption by dynamically trading their demand response resources [2, 3].

This new power system paradigm is usually referred as the smart grid (SG) as it allows the integration of all these "active" elements (including customers) in the physical system (grid) and new trading mechanisms and markets (mainly retail) [4, 5]. The concept is gaining importance as the solution for the future power system [6]. However, it is essential a proper operation and control, which is intended to be automatic through SG controllers, using the resources offered by generators and customers directly or through intermediate agents such as load aggregators or generation aggregators, usually referred as virtual power plants (VPP) too [7].

The requirements for building these SG are as follows:

- Reliability: they have to be more reliable than the traditional power systems. This has to be accomplishing providing them with self-healing capacity.

- Economic: they have to reduce the cost of the electricity by integrating all available generation and new technologies like control, storage, etc.

- Secure: they have to be secure against physical attacks or cyberattacks.

- Participation: they have to give the consumer more options to participate in energy or other services markets.

To build this type of SG application, it is necessary to deploy a large number of elements (control centres, smart metering system, renewable control system, storage elements, etc.) and operate them in a synchronised and cooperative way so that the above-mentioned objectives can be achieved. In this sense, the information and communication technologies may be the bottleneck for a correct implementation of the SG paradigm. One of the most relevant problems is to establish reliable two-way communication systems between the distributed elements of the SG.

Normally, the digitalisation and communication layer has been studied on a very abstract level [8]. However, the implementation of such layers requires to be deployed at the application level to overcome the practical problems of smart grids. Thus, it remains essential to clarify and select the different protocols that need to be applied in a smart grid. A summary of them can be seen in [4]. Nevertheless, it is important to note that power systems and their communication systems remain in constant growing and evolution, and new protocols or upgrades of them are constantly appearing [9]. Therefore, the protocols that can be selected should tend to maximise the functionalities and ensure the maximum standardisation possible. To do so, the maximum number of functionalities and interoperability can be selected in order to better adapt the flexible resources potentialities to the communication system. Moreover, the selection of standards also has to fulfil cybersecurity needs that enable a proper communication from the customer side to the system. Enabling a proper communication platform (such as internet) will remain essential to integrate distributed generation in the system.

Specifically, the transition to a more sustainable energy sector has gained importance in islands. Many islands are facing or will face severely the effects of climate change and are joining efforts to reduce their carbon footprints. However, their isolation and lower systems sizes represent a challenge to fully integrate volatile and intermittent renewables. Therefore, flexibility combined with renewable integration arises as a condition to make a viable transition from a fossil fuel-based system to a sustainable one. Thus, to capture all the benefits of renewable-based SG, islands needs to deploy reliable communication systems designed for their particularities. Moreover, island scenarios are analogue to the cases of electrification of isolated rural communities, which are widely studied [10–12]. These systems tend to rely on hybrid systems with renewables in island to overcome the lack of electricity [13]. The deployment of renewables in island mode requires better coordination and communications to ensure the security, reliability and viability of the system under the stochastic nature of renewable energy sources [14].

This chapter aims to design an architecture communication to provide the communication and interactions design between the different agents. Second, this architecture model is applied to the Santa Cruz Island in the Galapagos archipelago. Here, the implications to have a smart grid with proper communications are detailed. In this vein, a year simulation is performed to show the results of the proposed architecture and highlight both economic and environmental benefits arising from the deployment of smart grids with large share of flexible resources.

The rest of the chapter is structured as follows: Section 2 outlines the communication characteristics and architecture required to implement a smart grid on islands. Then, Section 3 describes the Galapagos case of study under the scenarios suggested and includes the discussion of the results. Finally, in Section 4, some conclusions are drawn.

2. Communication architecture for exploiting flexible resources

The exploitation of flexible resources to boost the integration of renewables requires a proper communication architecture. In this section, the solutions that need to be applied at distribution level to integrate demand response, electric vehicles and distributed generation have been stated. Special attention is also put into the deployment of smart metres to ensure an advanced metering infrastructure that enables the correct operation of the smart grid.

In order to design the communication solutions, four different types of interactions based on the characteristics of the different participants have been considered. The first three types are implemented without human intervention (full automation), while the last one requires human intervention. Before describing these interactions, it is needed to explain the concept of actor. In this context, an actor is defined as any agent involved in the operation of the power system but not taking part as a consumer, storage or generator.

1. "Actor-generator/storage" communication

In this case, the system operator or the VPP remotely operates a generator or a group of them in real-time to provide secondary and tertiary control reserve or voltage control to the power system.

Reviewing the existing communication protocols used to operate generators remotely, there are a lot of options but it is proposed to use the protocol "IEC 61850". It is expected that this protocol will be one of the most widespread solutions for integrating distributed energy resources (DER) in the distribution network when these resources are controlled by the DSO.

2. "Actor-consumer" communication

This communication is used to remotely change the electric consumption pattern of the customers that participate in a demand response programme in an orderly manner. In this regard, the exchange of data between an actor and a consumer includes different types of messages such as the transfer of an array of hourly prices (RTP), the customer acceptance of a DR event, the use of the direct load control (DLC), etc.

Regarding the proposed application protocol for this type of communication, the most widespread protocol related to demand response issues throughout the world is open ADR. One of its most important features is that it can be used to implement most of the demand response programmes existing nowadays. In particular, it is proposed to use the version 2 and the profile B due to the fact that it does not need to open any ports in customer's firewall improving security aspects and avoiding a lot of problems during the initial hardware setting.

As an alternative to open ADR, the Open Charge Point Protocol (OCPP) is also proposed for managing the electric vehicle charging using the version 2.0. This protocol is the most widespread one in this specific field. Regarding the management of EV charging, it is important to highlight that this kind of flexible resources can be considered as a load or an electric battery depending on if V2G option is implemented or not.

3. "Actor-actor" communication

Apart from customers and generators, the rest of agents employ centralised management systems to perform their main activities. In this vein, it is very common inside the smart grid paradigm that all actors need to exchange information between them in real time. In the proposed communication architecture for islands, there is only one case where the TSO uses the flexible resources for operating the power system. In this case, the DR server of the TSO has to be able to send signals to the DSO, who acts as a DR aggregator. To this end, it is proposed to use open ADR 2.0b, in spite of this protocol was not developed to resolve this specific goal.

4. "Human operator-system" communication

Although the degree of automation is becoming higher and higher, there are some tasks that has to be already carried out by human operators (e.g., initial setting of systems). Moreover, graphical user interface (GUI) trends to be implemented as a web application in order to reduce the cost for improving the compatibility with the different versions of operating systems and other required applications such as database software, communication libraries DLL, etc.

According to all this, it is proposed the Hypertext Transfer Protocol Secure (HTTPS) to exchange information between human operators and centralised management systems. This protocol is an extension of the hypertext transfer protocol that improves security thanks to the integration of the Transport Layer Security (TLS), which encrypts bidirectionally all the packets exchanged between client and server.

2.1 Application to distribution automation

The standards and systems for the communication at transmission level among the different components (energy management system, generators, substations and large customers) are relatively well established due to the relevance of their performance in the overall reliability of the power system. In fact, before the development of smart grids, most of the investment in the electricity sector was made on the reinforcement of the transmission grid and improvement of the associated energy management system.

Regarding the communication requirements at distribution level, a lot of changes are necessary to evolve the existing infrastructure and systems into smart grids. The idea of making distribution systems more automatic is not new. In this regard, the utilities, along the last 30 years, have been trying to reproduce the successful EMS in transmission systems into distribution. What really happened is that a set of independent applications were developed in the distribution control centres (managed by the Distribution System Operator, DSO) at the feeder and substation levels. The communications required for these applications were usually implemented by using vendor-dependent protocols and by implementing different and independent data networks in the past, but the current trend is to minimise the number of proprietary communication implementations and use open, standard-based specifications.

According to that, the protocol IEC 61850, for controlling network elements from SCADA, and ICCP-TASE-2, for exchange information between control centres, is proposed to successfully implement the required communication at the distribution. Nevertheless, recently, new requirements have arisen at the distribution level due to smart grid implementation. For example, the adequate integration of new distributed energy resources such as distributed generation or storage, even if managed by virtual power plants, requires new communication features in the existing or new protocols. Additionally, it is important to highlight that due to the security problems associated with the cited protocols, it is necessary to consider the recommendations of IEC 62351 in order to increase the resiliency of the proposed solutions.

In addition to the specific standards commented above, a set of suggested communication standards for distribution automation applications are summarised below, based on the recommendations of the International Electrotechnical Commission (IEC, [9]):

- IEC/TR 62357 service oriented architecture (SOA)
- IEC 61970 common information model (CIM)/energy management
- IEC 61968 common information model (CIM)/distribution management
- IEC 61850 power utility automation. The following subsections are a result of special interest:
    - IEC 61850-7-420—Communication systems for Distributed Energy Resources (DER)—Logical nodes
    - IEC 61850-7-500—Use of logical nodes to model functions of a substation automation system
    - IEC 61850-90-2—Use of IEC 61850 for the communication between control centres and substations

  - IEC 61850-90-6—Use of IEC 61850 for distribution feeder automation system
  - IEC 61850-90-7—Object models for photovoltaic, storage and other DER inverters
  - IEC 61400-25—Application of the IEC 61850 methodology for Wind turbines
- IEC 62351 security
- IEC 62056 data exchange for metre readings, tariffs and load control
- IEC 61508 functional safety of electrical, electronic, programmable electronic and safety related problems.
- IEC PAS 62746 interface between customer energy management system and power management system. Open automated demand response

2.2 Smart metres

The implementation of smart grids has to overcome all the problems that have encountered the utilities to enhance the distribution system. One key element in the communication chain in smart grids is the Advanced Metering Infrastructure (AMI) System that encloses all the elements to provide a reliable and secure communication to the most important partner in smart grids: the customer [15].

The main features of the AMI systems are:

- Smart metre, which is the frontier element with the customer, where bi-directional communication is implemented and where the smart metre may initiate a conversation with other agents in the smart grid. For example, it may send a black out (no supply voltage signal) before the metre goes off.
- The smart metres must be configurable in remote mode so the parameters of customer facility (for example, the protection settings) may be remotely changed without needing to go physically to the customer site.
- The smart metres may also be used as the gateway of customer network for other distribution management systems (Home Area Network, HAN, in residential customers).
- The smart metres have not been considered suitable for demand response implementation, as discussed later, because the possible delays and latencies introduced by other processes running in the metre.
- Another basic component, which is complementary to the AMI system, is the measurement data management (MDM) system that is an application responsible for the measurement data consistency and coherency checking as well as for the preparation of this data for further applications: commercial or technical.

According to what has been reviewed so far, it is clear that there are two main issues in developing smart grid communication structures and systems:

interoperability and cybersecurity. Interoperability may be warranted by using the adequate non-vendor protocols for communication, whereas cybersecurity may be improved considering the recommendations of standards devoted to this issue (e.g., IEC 62351).

2.3 Application to electric vehicles

Integration of EVs on the system is being widely studied due to the interest of decarbonizing transport [16]. In this regard, the interoperability and necessity to charge and discharge EVs is a topic of interest due to the large number of existent studies [17–19]. The proposed standards for charging electric vehicles are mainly three. on the one hand, open ADR is proposed to manage the EV charging of commercial and residential customers, especially in the case of wall-box charging stations due to the requirement of standardising the protocols for controlling loads at residential and small commercial customers. Their specifications are described in "Section 2.4". The IEC 61850 standard has also been studied for implementing the harmonisation of charging protocols of EVs [20].

Finally, "Open Charge Point Protocol" (OCPP) is proposed for public charging stations. This is an open and interoperable communication standard based on JSON over Websockets, including compression for data reduction what facilitates the interchange of information between EV centralised management system and the charging stations.

This protocol has been considerably improved in version 2.0 with extended functionality related to smart charging such as direct smart charging inputs form a centralised management system to a charging station or just with a local controller and supporting smart charging based on ISO15118.

OCPP 2.0 supports around 120 use cases that are integrated into 16 blocks according to different functionalities. These are completely described in [21] and are summarised below:

a. Security

b. Provisioning

c. Authorization

d. Local authorization list

e. Transactions

f. Remote control

g. Availability

h. Reservation

i. Tariff and costs

j. Metering

k. Smart charging

l. Firmware management

m. ISO 15118 certificate management

n. Diagnostics

o. Display message

p. Data transfer

Although there are several interesting functionalities for the deployment of smart grid, smart charging can be considered the most relevant for its impact on the feasibility of the proposed solution. The capacity of smart charging implies that a centralised EV charging management system gains the ability to influence the charging power or current of a specific electric vehicle or the overall energy consumption of an entire charging station during a period of time. Therefore, an external system has the possibility to set a charging profile as a limit of overall energy consumption of each charging station or group of them, what can be suitable to improve distribution network operation (network restrictions, balancing, loss reductions, etc.). This protocol is enabled to implement V2G applications.

2.4 Application to demand response

As commented above, the communication protocol “OpenADR” has been chosen since it has been identified as the best standard procedure for this application. Currently, it is the most used protocol to implement DR actions in power systems [22].

The OpenADR 2.0 standard includes two profiles:

- Profile A (OpenADR 2.0a). It has been designed for low-end embedded devices to support the basic services of demand response and markets.

- Profile B (OpenADR 2.0b). It has been designed for high-end embedded devices and it includes feedback to consumers as a response to events/data reports, present and futures.

The “OpenADR Alliance” has developed the procedures and necessary tools for the certification of products working with such protocol:

- The technical specifications for different profiles

- The documents establishing the characteristics that should be mandatory for each profile

- The testing plan; the certification documents

- Finally, the tools to perform the certification tests that may be used by third parties to check their products

A new concept, introduced in OpenADR 2.0, is the capacity to support two types of communication nodes: “Virtual Top Node” (VTN) and “Virtual End Node” (VEN).

The VTN represents a server that published and transmits OpenADR signals to final devices or other intermediate servers. The VEN is a client, a building energy management system (BEMS) or terminal device that accepts OpenADR signals from the VTN and responds to them. A final node may simultaneously be VTN and VEN. OpenADR signals are sent through standards based on internet protocols (IP) such as "Hyper Text Transfer Protocol" (HTTP) or "XML Measuring and Presence Protocol" (XMPP).

Information on a new event is contained in five parts:

- Event description: general metadata about the event
- Active period: starting time and total duration
- Event signal: interval data for the event
- Reference line for the event: interval data for the reference line
- Objectives: objective resources of the event (single VEN, defined VEN group, type of device, service area, type of resource, etc.)

Regarding security, open ADR uses a public key infrastructure (PKI) in order to provide:

- Authentication
- Confidentiality
- Integrity of transmitted data

There are two security levels between communications:

- Standard security: TLS with certificate interchange between server and client
- High security: it adds to standard security the utilisation of digital signatures for the XML data, so that the risk to reject information is reduced

Certified products should be provided with standard security, being optional the incorporation of high security.

According to the loads that need to be controlled and their communication needs, four different configurations for implementation have been considered:

- Connection to residential and small commercial consumers
- Connection to residential and small commercial consumers through a graphic user interface (GUI) on the cloud
- Connection to residential and small commercial consumers through an BEMS
- Direct connection to industrial and commercial consumers
- Aggregator model

Among other existing programmes, which may be implemented by means of open ADR, the following examples can be found:

- Residentia

    - "Save Power Day" (SPD)
    - "Summer Advantage Incentive" (SAI)
- Commercial and industrial
    - "Demand Bidding Program" (DBP)
    - "Summer Advantage Incentive" (SAI) also known as "Critical Peak Pricing" (CPP)
    - "Capacity Bidding Program" (CBP)
    - "Aggregated Managed Portfolio" (AMP)
    - "Base Interruptible Program" (BIP)
    - "Real Time Pricing" (RTP)

Apparently, the requirements for the communication between a DR requester (system operator, DR aggregator, etc.) and a DR provider are linked to the implemented type of DR programme. In this regard, a basic set of DR programmes was proposed to try to maximise the benefit of the potential flexibility of each type of flexible resource. They were classified into two groups depending on the degree of automation of the solution that is defined as:

- *DR programmes oriented to semi-automatic control systems*: customers receive a notification at least 1 hour before the starting time of a DR event, and they should set up their local control systems to implement the DR action on time. Some examples of this type of DR programmes are known as "Traffic lights" (CPP) more suitable for residential customers or "dynamic pricing (RTP)" for industrial customers. The strong point of this type of programmes is that there is no need to install additional hardware in the customer facilities, except from installing smart metres that it is necessary to hourly collect the energy consumption in real-time. However, the drawback is a lower percentage of customer who reacts to these kinds of signals and the lack of accuracy for implementing the response.

- *DR programmes for automatic control systems*: customer's resources are directly controlled by other agent, but there is a clear definition of how this control has to be implemented. In this regard, two types of signals can be received by the local control system to implement a DR event: a price signal or a power set point. Regarding the first type, it is proposed to use auto-DR programmes or fast DR programmes depending on the reaction time of each resources to implement the change of pattern that can be used to provide secondary reserve or primary reserve, respectively. On the other hand, it is proposed to use direct load control for large flexible resources due to the need of being more accurate in the implementation of DR events.

Finally, according to the proposed model, customers may have installed generation plants with an installed power lower than 100 kW with additional electric batteries that might be remotely controlled using open ADR as the explained flexible resources due to they have the same network conditions.

Regarding all these DR programmes, it is not recommended to use the smart metre as a gateway to communicate with local control systems and devices due to possible delays and latency problems in the communication network as mentioned above.

2.5 Application to distributed generation

Distributed generation can be defined as generation that is small, disperse and connected to the distribution grid [23]. Regarding VPP, it can be considered as the agent that coordinates small and disperse generation to perform as a single entity.

The kind of distributed energy resources depends on the installed power and has been characterised and the communication requirements settled according to the following three groups:

1. $\mathbf{P_{inst} < 10\ kW}$. In this case, there are no direct control actions on devices, being just necessary to install a smart metre.

2. $\mathbf{10\ kW \leq P_{inst} < 100\ kW}$. In this case, the VPP is responsible for the management, so that basic control (on/off) has been considered.

3. $\mathbf{P_{inst} \geq 100\ kW}$. For this kind of facilities, aggregation into a VPP is proposed in order to offer ancillary services, as well as voltage control to the TSO. This requires that distributed generators must be able to be controlled as if they were conventional generators, with PQ set points.

2.5.1 DG of $10\ kW \leq P_{inst} < 100\ kW$

Control requirements for this kind of generation, according to the conceptual design, are low. Therefore, just the possibility of disconnection in case it is necessary for the TSO has been considered. According to these requirements and considering that control orders would be submitted by the aggregators, the utilisation of the protocol open ADR 2.0b simple XMPP in Push mode has been considered for connection and disconnection. In Option 1, the installation of a multiprotocol communication gateway is proposed in order to connect to the SCADA (local control), or to a programmable logic controller (PLC). In Option 2, the local control system may have a SCADA able to receive messages according to the open ADR protocol by means of an additional software module. Both solutions consider the access to the Internet by means of any available technology, as shown in the figure:

In that case, devices to be installed would be the following:

Additionally, the installation of smart metres should be considered in the same rate as the communication gateway devices.

2.5.2 DG of $P_{inst} \geq 100\ kW$ (VPP)

Control requirements for this kind of generation would be similar to those expected from a conventional generator. In case of photovoltaic power plants, inverters will be responsible for incorporating this management option. Such configuration would allow omitting the SCADA, which may be substituted by a RTU able to use the **IEC-61850** protocol for monitoring and controlling the inverters by means of PQ signals.

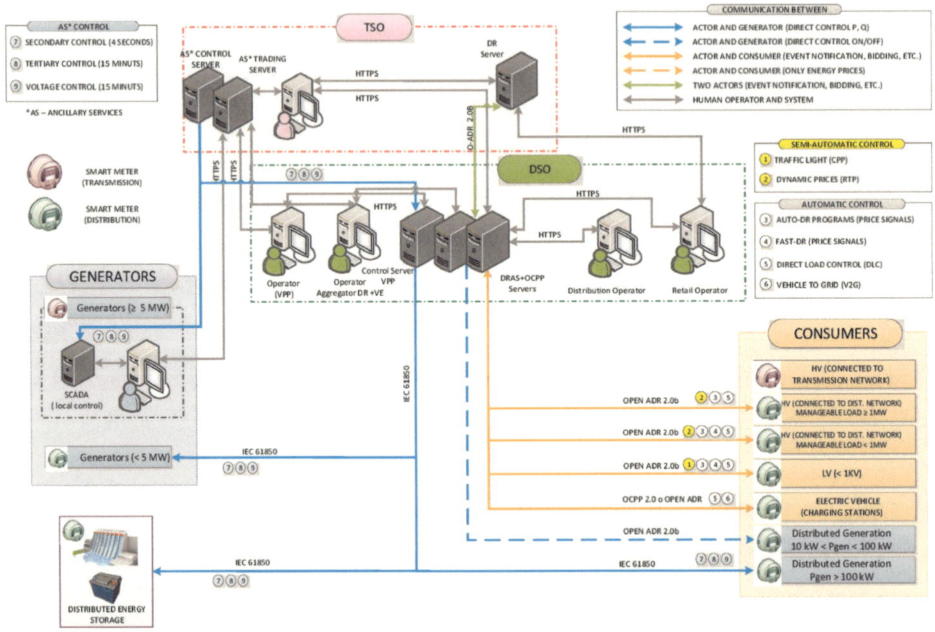

Figure 1.
Proposed communication architecture in Galapagos.

The same protocol IEC 61850 will be used to manage distributed storage such as batteries because in this particular case it is proposed to be performed by the DSO (**Figure 1**).

3. Case study of Galapagos

The proposed architecture is applied to the Santa Cruz Island, located in the Galapagos archipelago (Ecuador). These 21 islands are located 1000 km West of Ecuador's mainland (**Figure 2**). Approximately 26,000 people inhabit them; being Santa Cruz Island the most populated one of them. Notwithstanding its bio reserve protected status and an increasing concern about its preservation, tourism keep growing. From 1990 to 2015, visitors have doubled every 5 years [24], increasing the pressure over natural resources, especially electrical power demand. In this context, a transition to a renewable and sustainable energy mix for the archipelago is crucial to safeguard it [25]. Thus, the Ecuador government launched a Zero Fossil Program for Galapagos in 2007 to achieve a zero emission goal [26]. Nevertheless, this programme raises several technical difficulties due to the small size of the archipelago's mini grids and their low capacity to absorb and integrate renewables. The implementation of a smart grid architecture in Galapagos is aligned with one of the main energy and sustainability challenges that small islands and systems are facing currently, energy dependence, reliability and climate change [27]. Moreover, it boosts the efforts that are being performed to improve the current efforts to integrate renewables in small islands [28, 29].

3.1 Electricity system of Santa Cruz

Santa Cruz operates as an isolated mini grid coupled with the Baltra Island, where the island's airport is located. The island's electricity is mainly generated

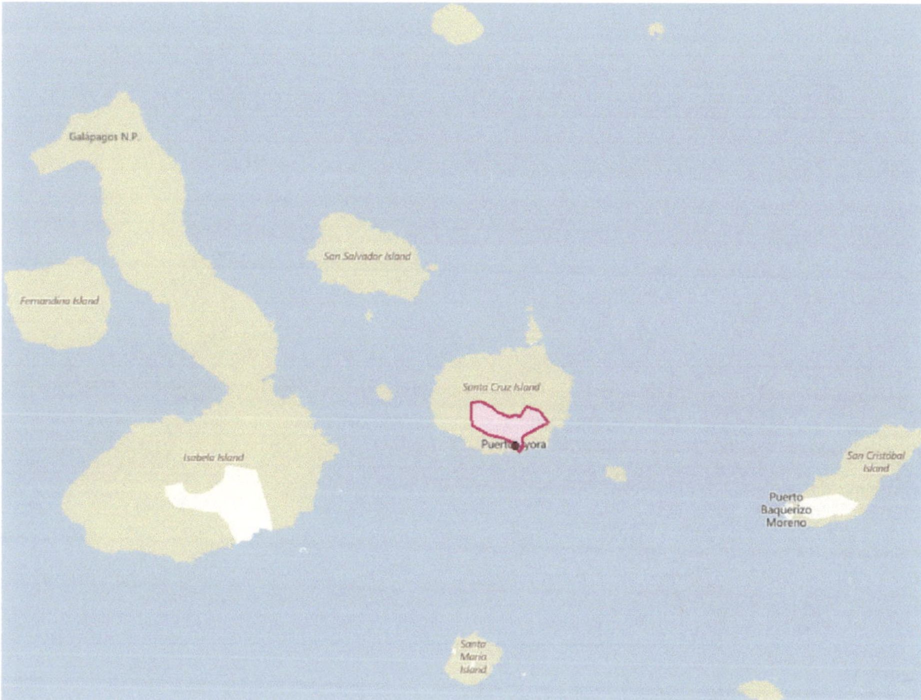

Figure 2.
The island of Santa Cruz in Galapagos.

with diesel that has to be imported from mainland. Currently, 2.25 MWp of wind power and 1.5 MWp of solar PV are installed in the island. However, both of them combined did not cover more than 12% of the total electricity demand in 2017. The main consumption occurs in the urban area and has the potential to become flexible and integrate renewables on it. The existing distribution network can be seen in **Figure 3**.

Due its geographical isolation, Galapagos and particularly Santa Cruz represent an extreme case to analyse the viability of the proposed architecture and its replicability. Thus, the transformation of Santa Cruz into a smart grid as the one proposed in this chapter offers several benefits at a reduced cost. The proposed architecture can be easily implemented in Santa Cruz, facilitating renewable energy integration and improving the system performance.

3.2 Simplified model

In order to assess the potentials of the smart grid, we propose a simplified *ex post* energy balance with historical and modelled data. The first requirement in the analysis is to fulfil at any time step the energy balance presented.

$$P_t^D + P_t^L = \sum_g P_{gt} + \sum_g \Delta t * f\mathrm{r}C_{\mathrm{gt}} \quad (1)$$

The generation in the smart grid with high penetration of renewable energy source scenario, the performance of the existing wind turbines and the solar farm; and the measured wind speed at hub height are used to linearly model the performance of the newly integrated renewable energy sources. Thus, the wind has been modelled with wind speed levels at 10 m. These have been corrected to get them at

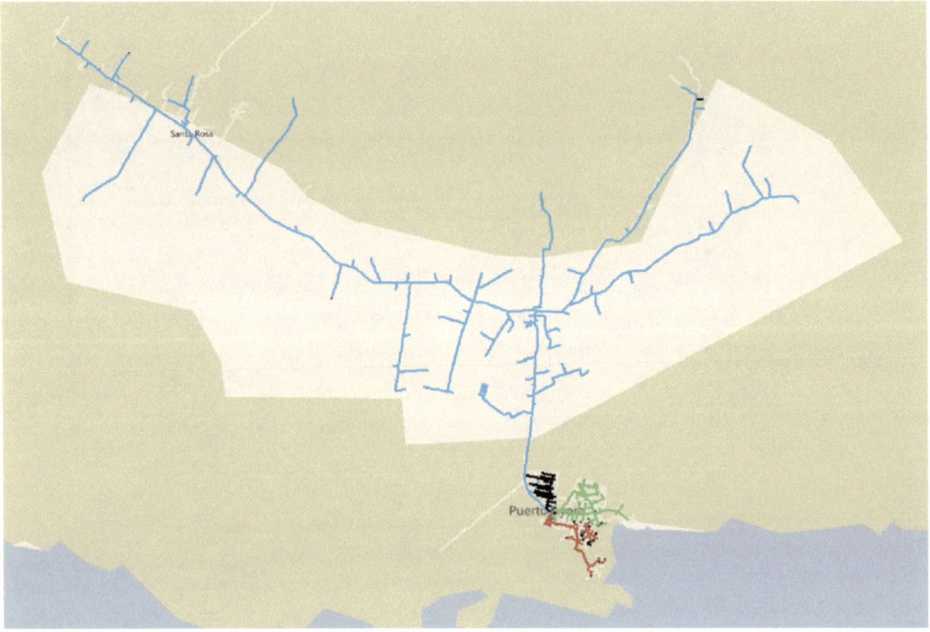

Figure 3.
Medium voltage distribution system in Santa Cruz [30].

80 m with a logarithmic correlation and then the wind production curve has been applied. The solar PV power has been calculated as a correlation of the actual production and a coefficient to increase the PV capacity. And at any time step, the system has to be inside the maximum and minimum capacity levels.

$$P_{gt}^{W} = \frac{1}{2}\rho\eta_t R^2 \pi u_t^3 \tag{2}$$

$$\underline{P_g} \leq P_{gt} \leq \overline{P_g} \tag{3}$$

The thermal capacity has been modelled as the difference between demanded electricity and losses. Thus, the existing thermal generator acts in a flexible manner to adjust generation to demand as it currently does.

$$P_t^D + P_t^L - \left(\sum_g P_{gt}^{PV} + \sum_g P_{gt}^{W} + \sum_g \Delta t * fr_{gt}\right) = \sum_g P_{gt}^{T} \tag{4}$$

The flexible resources such as demand response and electric vehicles have been characterised as batteries. This is a common model simplification [8, 9]. The flexible resources have a state that is among a maximum and minimum capacity and is represented by its state in the prior time step plus the charge or discharge ratio during the time period. Moreover, the charge and discharge ratios also have limits. Thus, these elements are modelled as:

$$\mathrm{SOfr_{gt}} = \mathrm{SOfr_{g(t-1)}} + \Delta t * frC_{gt} \tag{5}$$

$$\underline{SOfr_h} \leq \mathrm{SOfr_{gt}} \leq \overline{SOfr_h} \tag{6}$$

$$\underline{frC_h} \leq \mathrm{frC_{gt}} \leq \overline{frC_h} \tag{7}$$

No network constraints have been considered since it does remains out of the scope of this work. However, the costs associated to the increase on the distribution lines due to the new generation have been included in the installation costs.

The environmental analysis has been performed according to the emissions associated with electricity generation. Thus, the emissions have been calculated with emission factors associated with each type of fuel.

$$\mathrm{GHGe} = \sum_{i} \mathrm{EF_i} * \mathrm{q_i} \tag{8}$$

Finally, the analysis of the scenarios has been done with two basic economic indicators, the net present value (NPV) and the return over investment (ROI):

$$\mathrm{NPV} = \mathrm{Inv} + \sum_{y} \frac{CF_y}{(1+j)^y} \tag{9}$$

$$\mathrm{ROI} = \frac{\mathrm{NPV}}{\mathrm{Inv}} \tag{10}$$

3.3 Description of the scenarios

To study the benefits that arise from the smart grid deployment, three scenarios have been analysed and compared: business as usual, current status with smart grid and smart grid with high penetration of renewable energy sources.

First, the current situation of the system is taken as the "Business As Usual" (BAU) scenario. The system performance in 2017 is taken as the base for this scenario. As mentioned above, this scenario is characterised by 1.5 MW of solar PV, 2.25 MW of wind power and a large reliance on thermal generation to cover almost 90% of the electricity consumption. Currently, some wind production is curtailed due to the low flexibility of the system.

The second scenario is named "Smart Grid". The smart grid scenario is able to integrate all the wind generation that is currently curtailed thanks to the flexibility given by the system installed in this scenario and the better information channels implemented. This flexibility capacity is obtained from demand response resources, the VPP and the management of the EV charging stations installed. Their capacities have been estimated has percentages of the total demand and the study of the flexibility options that arise from them. The aggregator manages these programmes, which provides a flexibility that afterwards is sold to the VPP. Then the VPP manages this flexibility to provide firm power to the grid. Regarding the capital cost required to install the smart grid, the prices here summarised have been obtained from [30]. Different international smart systems and automation companies proposed solutions. **Table 1** presents a summary of the average prices communicated by them.

The third scenario is named "Smart Grid + Renewables". The same smart grid as the one proposed in the second scenario is implemented. Moreover, the integration of new renewable energy capacity has been optimised to obtain the maximum share of renewable energy generation without curtailment. This capacity can be integrated due to the increase in flexibility associated with the smart grid. Demand response programmes and an optimisation of the system performance allow the system to largely increase the share of renewables in the electricity mix. For this scenario, the capital cost of the newly installed generation has been retrieved from 2017 IRENA's world averages [31]. Prices are assumed to be 1.388 and 1.477 $/kW for solar PV and onshore wind, respectively. With the current grid infrastructure and the flexibility that the smart grid systems provide, the Santa Cruz Island could

Element	Cost
Smart metering	$ 318,200
Demand side management[1]	$ 950,000
Smart grid control	$ 150,000
Virtual power plant	$ 450,000
Total	$ 1868,200

[1]*The DSM includes all the infrastructure and equipment to implement DR.*

Table 1.
Smart grid costs.

Technology	Cost
Smart grid	$ 1868,200
Solar PV	$ 4,302,800
Wind	$ 3,323,250
Total	$ 9,494,250

Table 2.
Smart grid with high penetration of renewable investment cost.

absorb renewable energy generation from up to 4.6 MW of solar PV and 4.5 MW of wind power. Thus, 3.1 MW of solar PV and 2.25 MW of wind power could be installed. According to the abovementioned prices, the total capital cost required for this scenario is (**Table 2**):

The analysis and modelling of the scenarios has been performed to fulfil power and energy balances in 10 minutes time steps. The simulations have taken into account the real data from Santa Cruz Island. The data used has been electricity demand and solar PV, wind and thermal generation data. Whenever wind production has been curtailed, the theoretical power values of the wind turbine have been calculated with the power curve characteristic and the 10-minute wind speed average at the wind turbine's hub height data.

Finally, regarding the economic analyses, different data from the country has been used. Despite the diesel end-price in Ecuador is subsided, a non-subsidised price of $1.08 per litre is used to analyse the economic benefits of the implementation. This price is used according to the local ratio where the subsides cover 75% of the mentioned fuel, and the current prices at April 2018 are $0.27 per litre in Ecuador and $1.04 per litre as the World's average [32]. The cash flows have been calculated for the smart grid and smart grid + renewables scenario as the associated savings of not using the fuel. For both scenarios of smart grids, the parameters assumed to evaluate the investment are as follows: the lifetime period has been considered 10 years; the cash flows have been assumed constant during this period; the economic benefits are based on a yearlong energy balance simulation and the interest rate used has been 8%. For the environmental analysis, an emission factor of 2.67 kg CO_2 per litre of diesel has been used after the U.S. EPA [33].

3.4 Discussion

In order to assess the potentialities of each of the proposed scenarios, the business as usual scenario is presented in the next **Figure 4**. As it can be seen, the demand in mainly covered by thermal generation as renewable generation just covers about 10% of it. Moreover, as it has been stated part of the wind power generation was curtailed.

The smart grid scenario deploys the above-explained infrastructure. This new communication infrastructure improves the efficiency of the system and the renewable integration. The smart grid with flexible resources allows improving the capacity factor of the wind production from 13 to 26.3% with an improved information system that allows managing the energy production and optimise the renewable output. The improvement and digitalisation of the current grid infrastructure will allow a reduction of 24 kTnCO_2. The required investment for the implementation of the proposed smart grid architecture would be $ 1868 million, which will be rapidly recovered due to the economic savings obtained from fuel savings. The payback would be lower than 2 years, the net present value will amount benefits up to $3.7 million. This shows how the investment would be not only profitable but highly valuable in economic terms with a Return Over Investment (ROI) of 197.78% in 10 years. Moreover, the installation of the smart grid systems will pave the way for future investments and newly renewable energy generation (**Figure 5**).

In the third scenario, new solar PV and wind capacity are installed in order to maximise the share of renewable energy production. The smart grid architecture provides the needed flexibility to integrate generating facilities characterised by their stochastic nature. This integration allows the Santa Cruz Island to reduce more than 46% of the current emissions associated to electricity production. Besides the environmental benefits, the economic benefits of this action would be positive too. The installation of new capacity requires a high initial investment, which will be recovered in less than two and a half years. Moreover, due to the large diesel demand reduction, the benefits associated with this project would add to $12.4 million. Thus, being a profitable investment with and ROI of 130.15% (**Figure 6**).

To sum up the scenarios, the following figure shows a comparison among the energy balances of the presented three scenarios. As it can be seen, the current situation shows a system with large diesel consumptions associated to thermal generation while renewable energy has to be curtailed. The smart grid scenario provides an improved picture; the communication system associated with flexible

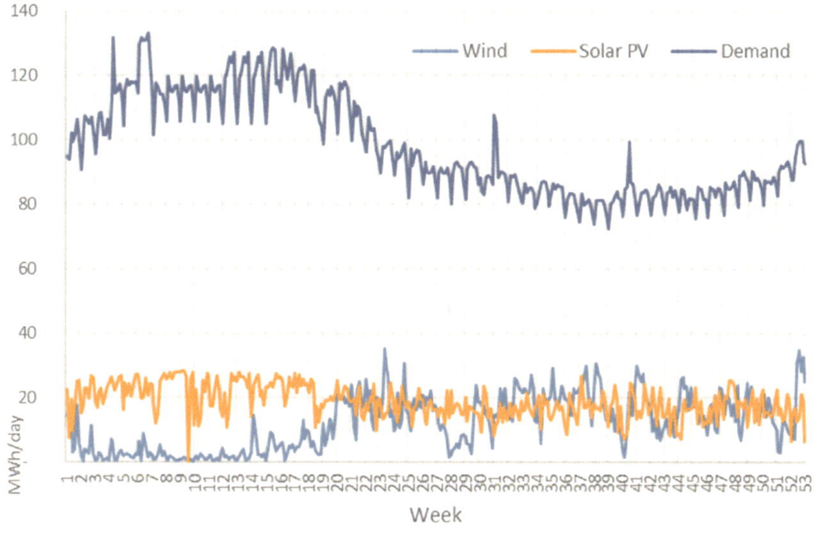

Figure 4.
Daily energy balance 2017. BAU scenario. Own elaboration based on [30].

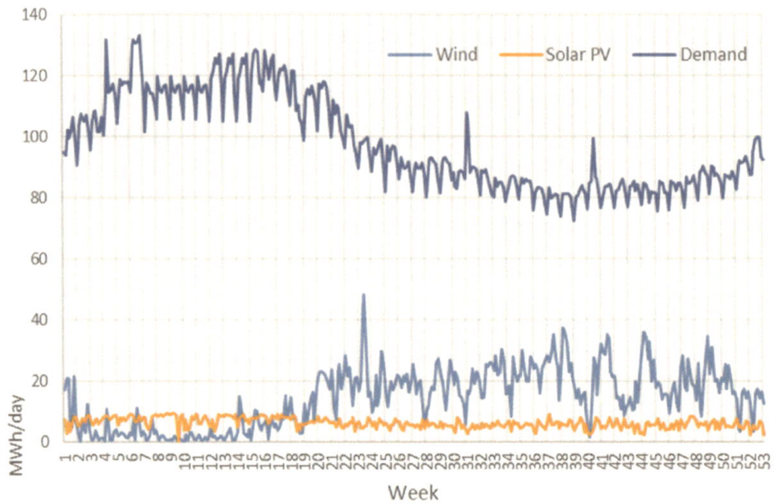

Figure 5.
Daily energy balance 2017. Smart grid scenario.

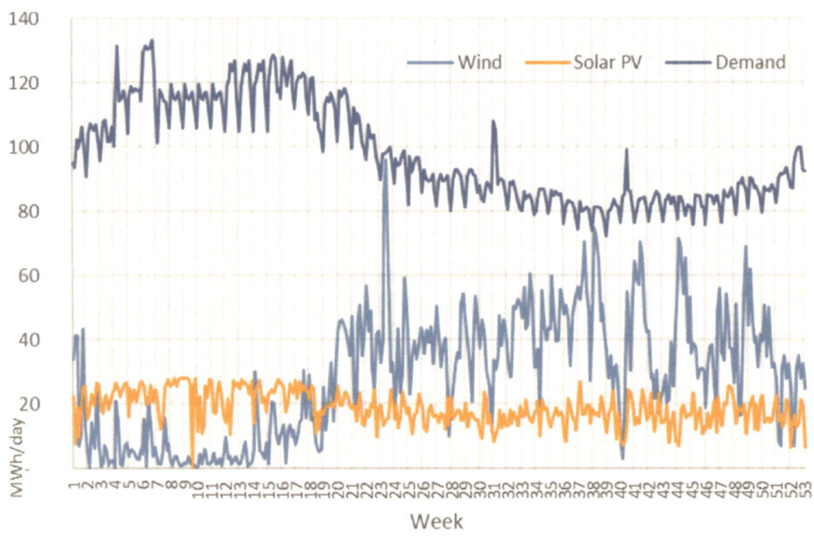

Figure 6.
Daily energy balance 2017. Smart grid + renewables scenario.

demand allows a total integration of the renewable generation. Finally, the third scenario depicts an energy mix with a share of renewables that accounts almost up to half of the total generated energy (**Figure** 7).

The variations between scenarios in renewable integration are associated with economic expenses and environmental impacts. In the BAU scenario, the Ecuadorian government is spending approximately $9.3 million a year just in diesel consumption. Moreover, this scenario is far away from the Galapagos Zero Emissions target with more than 224 $kTnCO_2$ emitted during 2017. Even though the 2.25 MW of wind and 1.5 MW of solar PV that are currently installed, the share of renewable energy generation in 2017 accounted just up to 11.2%. However, the renewable energy production could be optimised with the proposed smart grid implementation and the installation of new renewable capacity.

Regarding flexible resources, the residential, EV and commercial demand response centralised in the role of the aggregator allow the system to manage up to

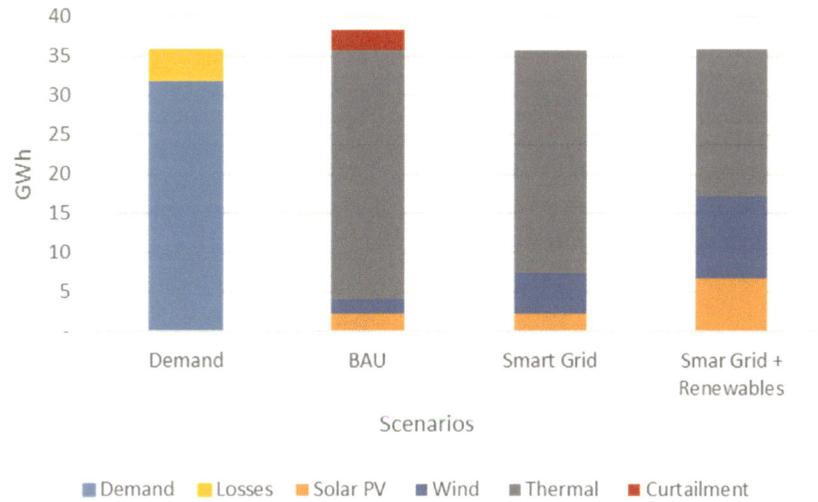

Figure 7.
Scenario's energy balance.

634 MWh/year. From this amount, EVs provide 50% of the flexibility. Commercial consumers provide 40% of the flexibility. The commercial DR is used 500 hours a year with an average flexibility of 250 kW. On the other hand, residential DR is only used during 85 hours a year, being these ones the most critical for renewable integration. Thus, commercial flexibility is used 10.5% of the time while the residential flexibility is used just 1% of the time. This will generate lower distortions to actual residential consumption patterns. Something may help the social acceptance of the project.

The results of the three scenarios proposed are summarised in **Table 3**:

As it can be seen, the payback of investing in the proposed architecture would have a payback lower than 2 years and a ROI of 198% while reducing 24 k $TnCO_2$.

Element	Scenario		
	Business as usual	Smart grid	Smart grids + renewables
Solar PV (MW)	1.50	1.50	4.60
Wind energy (MW)	2.25	2.25	4.50
DR commercial (MW/MWh)		1.0/4.1	1.6/4.1
DR residential (MW/MWh)		0.5/1.0	0.8/1.0
EV (MW/MWh)		0.9/1.6	0.9/1.6
RE integration (%)	11.2	20.3	47.5
Diesel consumption (m^3)	8609	7688	4991
Emissions ($kTnCO_2$)	224.7	200.6	130.3
Investment (M$)	—	1.87	9.49
Annual savings (k$)	—	0.99	3.91
Payback (years)	—	1.90	2.43
NPV (M$)	—	3.70	12.36
ROI (%)	—	197.8	130.2

Table 3.
Scenarios' results.

The inclusion of new agents as the VPP and the DR aggregator provides to the power system more flexibility that allows a larger integration of renewables that could increase from 11.2 to 47.5%. Both alternative scenarios present environmental and economic benefits. On the one hand, the implementation of the smart grid with flexible resources has a lower payback, while the scenario smart grid + renewables presents more benefits on the long rung due to the significant reduction of the diesel consumed. This latter scenario shows also a larger energy security due to the lower needs in diesel transportation and storage to the islands.

4. Conclusions

This paper presents a communication architecture that allows efficient communication flows to fully exploit the potential of flexible resources in order to accommodate larger integration of renewable energy sources. Based on the different protocols and features of smart grids, the necessities have been characterised, specially the inclusion of flexible resources such as demand response programmes, the electric vehicle, storage and distributed generation.

The communication architecture selected for Galapagos has proven to be a feasible solution in both technical and economic terms. The protocols selected have focused on maximising the functionalities due to their higher levels of development. First, the IEC-61850, following the security recommendations of IEC-62351, was selected as the main standard to control distributed energy resources (connected to the distribution grid) considering the use of distributor's private data network. Second, open ADR 2.0 was selected to ensure a secure and reliable implementation of the demand response among consumers in order to control loads and small renewable generation inside the customer facilities. Finally, the OCPP was selected as the protocol to manage the charge/discharge of EV batteries due to its large catalogue of functionalities. These two solutions are proposed to be implemented using the Internet network for reducing the required investment.

The ecological value of Galapagos is undoubtable and a transition to a sustainable energy system is mandatory. However, their particular orography and isolation generate to Galapagos Islands' grids a difficult challenge to achieve renewable energy integration. For this reason, a case study in the island of Santa Cruz has been studied. The implementation of a smart grid with the proposed architecture in Santa Cruz can help the island to perform an effective transition to become zero fossil fuel. A 1-year simulation in the third scenario highlights the importance of flexible resources to integrate renewable generation. Nevertheless, the results show that the implementation of a smart grid could not only immediately improve Santa Cruz power system emission factor and efficiency but will also allow future integrations of renewables.

As a conclusion, the proposed smart grid architecture proves to be valid and efficient in both economic and environmental terms. The inclusion of flexible resources on the system proves to be a valuable asset to help integrating renewables. This can be done thanks to the aggregation of the individual responses and the exploitation of electric vehicles, thus showing the importance of managing the flexibility of the system.

Acknowledgements

The authors would like to thank the Ministry of Ecuador of Electricity and Renewable Generation and the utility Electrogalapagos for their valuable help and comments.

Conflict of interest

"The authors declare no conflict of interest."

Nomenclature

Indexes

t	index of time periods
g	index of generation elements
i	index of fuel type
h	index of flexibility elements
y	index of years

Parameters

ρ	air density
R	radius
$\overline{P_g}$	maximum generation capacity of the generator g
$\underline{P_g}$	minimum generation capacity of the generator g
$\underline{SOfr}_h$	minimum state of charge of the flexible resource h
$\overline{SOfr_h}$	maximum state of charge of the flexible resource h
$\underline{frC}_h$	maximum discharge capacity of the flexible resource h
$\overline{frC_h}$	minimum charge capacity of the flexible resource h
EF_i	emission factor of the fuel type i

Variables

$GHGe$	green house gas emissions of the scenario
P_t^D	electricity demand during the time period t
P_t^L	electricity losses during the time period t
P_{gt}	electricity generated by generator g during the time period t
P_{gt}^{PV}	electricity generated by PV generator g during the time period t
P_{gt}^{W}	electricity generated by wind generator$g$ during the time period t
P_{gt}^{T}	electricity generated by thermal generator g during the time period t
$SCfr_{ht}$	state of charge of the flexible resource h during the time period t
frC_{ht}	charge/discharge power of the flexible resource h during the time period t
q_i	quantity consumed during the year of the fuel type i
CF_y	cash flow during the year y

Author details

Javier Rodríguez-García, David Ribó-Pérez*, Carlos Álvarez-Bel and Manuel Alcázar-Ortega
Universitat Politècnica de València, Institute for Energy Engineering, València, Spain

*Address all correspondence to: david.ribo@iie.upv.es

References

[1] Helm D. Burn Out: The Endgame for Fossil Fuels. New Haven, Connecticut, USA: Yale University Press; 2017

[2] Niesten E, Alkemade F. How is value created and captured in smart grids? A review of the literature and an analysis of pilot projects. Renewable and Sustainable Energy Reviews. 2016;**53**: 629-638

[3] Burger S, Chaves-Ávila JP, Batlle C, Pérez-Arriaga IJ. A review of the value of aggregators in electricity systems. Renewable and Sustainable Energy Reviews. 2017;**77**:395-405

[4] NIST. Framework and Roadmap for Smart Grid Interoperability Standards, Release 3.0; 2014

[5] Smart Grid Coordination Group. Smart Grid Reference Architecture. CEN-CENELEC-ETSI. Brussels; 2012

[6] Ipakchi A, Albuyeh F. Grid of the future. IEEE Power and Energy Magazine. 2009;7(2):52-62

[7] Rodríguez-García J et al. Novel conceptual architecture for the next-generation electricity markets to enhance a large penetration of renewable energy. Energies. 2019; **12**(13):2605

[8] Kaitovic I, Lukovic S. Adoption of model-driven methodology to aggregations design in smart grid. In: 2011 9th IEEE International Conference on Industrial Informatics. 2011. pp. 533-538

[9] Sundararajan A et al. A survey of protocol-level challenges and solutions for distributed energy resource cyber-physical security. Energies. 2018;**11**(9): 2360

[10] Muh E, Tabet F. Comparative analysis of hybrid renewable energy systems for off-grid applications in southern Cameroons. Renewable Energy. 2019;**135**:41-54

[11] Miao Y, Jia Y. Hybrid decentralised energy for remote communities: Case studies and the analysis of the potential integration of rain energy. Journal of Sustainable Development of Energy, Water and Environment Systems. 2014; **2**(3):243-258

[12] Chakravorty U, Emerick K, Ravago M-L. Lighting up the Last Mile: The Benefits and Costs of Extending Electricity to the Rural Poor; 2016

[13] Mandelli S, Barbieri J, Mereu R, Colombo E. Off-grid systems for rural electrification in developing countries: Definitions, classification and a comprehensive literature review. Renewable and Sustainable Energy Reviews. May 2016;**58**:1621-1646

[14] Hubble AH, Ustun TS. Composition, placement, and economics of rural microgrids for ensuring sustainable development. Sustainable Energy, Grids and Networks. 2018;**13**:1-18

[15] Benzi F, Anglani N, Bassi E, Frosini L. Electricity smart meters interfacing the households. IEEE Transactions on Industrial Electronics. 2011;**58**(10):4487-4494

[16] Geels BFW, Sovacool B, Schwanen T, Sorrell S. Accelerating innovation is as important as climate policy. Science. 2017;**357**(6357):1242-1244

[17] Morales DX, Besanger Y, Sami S, Alvarez Bel C. Assessment of the impact of intelligent DSM methods in the Galapagos Islands toward a smart grid. Electric Power Systems Research. 2017; **146**:308-320

[18] Hannan MA, Lipu MSH, Hussain A, Mohamed A. A review of lithium-ion

battery state of charge estimation and management system in electric vehicle applications: Challenges and recommendations. Renewable and Sustainable Energy Reviews. 2017;**78**: 834-854

[19] Jha M et al. Intelligent control of converter for electric vehicles charging station. Energies. 2019;**12**(12):2334

[20] Hussain SMS, Ustun TS, Nsonga P, Ali I. IEEE 1609 WAVE and IEC 61850 standard communication based integrated EV charging management in smart grids. IEEE Transactions on Vehicular Technology. 2018;**67**(8): 7690-7697

[21] Open Charge Alliance. Open Charge Point Protocol 2.0; 2018

[22] OpenADR Alliance. Open ADR 2.0 Profile Specification; 2012

[23] Ackermann T, Andersson G, Söder L. Distributed generation: A definition. Electric Power Systems Research. 2001;**57**(3):195-204

[24] Tourtellot J. Galápagos Tourism Backfires. National Geographic Blog; 2015

[25] Quinn MS. Driving Forces and Barriers for a Sustainable Energy Mix in Fragile Environments: North-South Perspectives. Cham: Springer; 2018. pp. 21-32

[26] Ministerio de Electricidad y Energía Renovable. Cero Combustibles Fósiles Galápagos; Ministerio de Electricidad y Energía Renovable. 2018. Available from: http://www.energia.gob.ec/cero-combustibles-fosiles-en-galapagos-2/ [Accessed: April 24, 2018]

[27] Cole P, Banks G. Renewable energy programmes in the South Pacific—Are these a solution to dependency? Energy Policy. 2017;**110**:500-508

[28] Blechinger P, Cader C, Bertheau P, Huyskens H, Seguin R, Breyer C. Global analysis of the techno-economic potential of renewable energy hybrid systems on small islands. Energy Policy. 2016;**98**:674-687

[29] Nepal R, Jamasb T, Sen A. Small systems, big targets: Power sector reforms and renewable energy in small systems. Energy Policy. 2018;**116**:19-29

[30] Álvarez C, Pesantez-Sarmiento PA Rodríguez-García J, Alcázar-Ortega M, Carbonell-Carretero JF. Análisis para la implementación del programa de redes inteligentes en Ecuador: Diseño conceptual y aplicación a plan piloto. 2016

[31] IRENA. Renewable Power Generation Costs in 2017; 2018

[32] GlobalPetrolPrices. Los precios del diesel en todo el mundo, 23-abr-2018| GlobalPetrolPrices.com; 2018. Available from: https://es.globalpetrolprices.com/diesel_prices/ [Accessed: April 24, 2018]

[33] US-EIA. How much carbon dioxide is produced when different fuels are burned? U.S. Energy Information Administration (EIA). 2018. Available from: https://www.eia.gov/tools/faqs/faq.php?id=73&t=11 [Accessed: April 25, 2018]

Chapter 8

The Strategies of EV Charge/ Discharge Management in Smart Grid Vehicle-to-Everything (V2X) Communication Networks

Ujjwal Datta, Akhtar Kalam and Juan Shi

Abstract

Electric vehicles (EVs) are at the forefront of the revolutionized eco-friendly invention in the transportation industry. With automated metering infrastructure (AMI) communications in houses, smart EV charging stations, and smart building management systems in smart grid-oriented power system, EVs are expected to contribute substantially in overall energy planning and management both in the grid and the customer premises. This chapter investigates and provides an in-depth analysis on the charge/discharge management of EV in vehicle to home (V2H), vehicle to drive (V2D), vehicle to vehicle (V2V), vehicle to grid (V2G), vehicle-to-building (V2B), and grid to vehicle (G2V). The planning and control of energy exchange of EV is the main focus considering EV availability in multiple places during the daytime and in the evening. Indisputably, EV participating in V2G or V2H affects the state of charge (SOC) of EV battery, and therefore proper scheduled charge/discharge plan needs to be embraced. The structures of EV in various operation modes and approaches are presented for implementing the energy planning and charge/discharge management of EV in different operation modes. The simulation results demonstrate the effectiveness of the proposed charge/discharge management strategy and regulation of EV SOC in accordance with the energy management plan of EV owner.

Keywords: electric vehicles (EVs), vehicle to home (V2H), vehicle to grid (V2G), vehicle to vehicle (V2V), grid to vehicle (G2V), vehicle to building (V2B), vehicle to drive (V2D), charge/discharge management, SOC constraints

1. Introduction

With the increasing concern of greenhouse gas (GHG) emissions, many attempts have been suggested and already placed in action for clean energy practice. Electric vehicles (EVs) are one of the revolutionized modern technologies in the transportation industry that has drawn the greater attention of market investors, governments, and customers. EVs are considered to have a significant contribution in reducing GHG emissions. However, in order to access the EVs impact on the way to clean energy and climate change, this requires the appropriate transformation and deployment of economy regulations by the governments [1, 2].

In recent years, the developments in different areas of EV technologies are focused on scheduled charging problem [3], minimizing fuel consumption with a variable driving schedule [4], increasing battery charger efficiency [5], and energy incentive policy for EV [6]. However, with the development of smart grid concept and the availability of bidirectional charging facility, EVs are considered to play a diverse role that can bring several benefits in the smart city power grid [7]. This will provide the opportunity for EV to enact as a power source and exchange energy with the grid for delivering multiple services [8]. The pricing-controlled EV charging in a smart city can reduce the additional burden on the network during peak time and subsequently provide economic benefit [9]. The study in [10] concluded that an optimized investment and operation is imperative to achieve a considerable amount of economic advantage. However, an extensive survey analysis by the study in [11] argued that profitability depends on several factors such as types and prices of regulation services and market regulations. The increasing penetration level of intermittent renewable energy sources has a positive impact on reducing GHG emissions. However, the grid voltage, frequency, and power quality are adversely affected due to the nature of their variable power output. The power grid needs to compensate and balance power differences to maintain a stable grid operation. EVs can provide significant support in compensating such power imbalances.

With the increased energy capacity of EV batteries, the energy requirement of the grid, charging/discharging flexibility of EVs, and concepts of vehicle to home (V2H), vehicle to grid (V2G), and vehicle to vehicle (V2V) have become more desirable for grid-connected operation. The V2H can provide multiple energy services of a house through smart building management systems during peak periods [12] and reliability enhancement when a load shedding occurs [13]. In V2H, EV can be coordinated for flexible load scheduling [14] and optimal energy management with photovoltaic generation [15], thus increasing the benefit of implementing V2H [16]. In V2G/V2V, EV can be exploited to minimize network variations and impact positively in the grid operation [17]. Few experimental studies have also been carried out to show the effectiveness of EV in different operation modes [18, 19]. Nevertheless, in order to participate in providing ancillary services, an efficient charging/discharging strategy for EV must be considered. Regardless of operation modes, smart energy management of EV can ensure reduced energy consumption from the grid and thus provide direct economic benefit to the customers [20].

This chapter proposes an in-depth analysis and discussion on EV energy management in V2D, V2H, V2V/V2G/V2B, and G2V. The key objective is to describe various approaches in EV battery charge/discharge control strategies in different operation modes, including the modeling of charge/discharge management methods, types of ancillary services, and feedback regulations to provide the aforementioned services. An energy pricing plan is included in charge/discharge of EV for "vehicle to everything (V2X)" services which is able to compute the economic advantage of providing V2X services and G2V.

The rest of the chapter is organized as follows. Section 2 presents the concept of grid services and existing typical EV energy capacity. The structure of EV in different operation modes is discussed in Section 3. Feedback regulation strategy in implementing V2X technology is described in Section 4. The management of EV battery charging/discharging in numerous operation modes is explained in Section 5. Simulation studies are demonstrated in Section 6, and the conclusion is drawn in Section 7.

2. Energy capacity of EV and V2X concept

The battery energy capacity is one of the most important factors in planning EV for V2X services. This is particularly important as depleting EV battery for V2X without an appropriate plan may threaten EV availability for V2D. The capacity of currently available battery ranges from a few kWh to 100 kWh [21]. The battery capacity of Ford Focus electric 2018 car is 33.5 kWh, Nissan LEAF 2018 is 40 kWh, Chevy Bolt 2018 is 60 kWh, Tesla Model 3 is 80.5 kWh, and Tesla Model X or Tesla Model S is 100 kWh [21]. Although battery capacity is not the only parameters that decide EV for V2X, it firmly indicates the distance that can be covered before the next charging event is required. In some cases, large EV battery capacity does not necessarily mean long distance coverage such as Tesla Model X has lower mileage (383–475 km) with higher kWh than Tesla Model 3 (499 km) [22]. Therefore, EV for V2X can be planned if sufficient state of charge (SOC) of EV battery is available at the end of the journey and the estimated energy requirement for next journey. Nevertheless, these ratings are fairly indicative assumptions, and actual energy consumptions per traveling distance may vary in real condition depending on several factors such as battery age, maintenance of the EV, nature of driving, number of passengers, weather conditions, etc.

In order to charge/discharge, i.e., load/energy supplier, EVs are plugged in at home or in a charging station through a charging outlet. EVs come with onboard unidirectional or bidirectional battery charger and range from various charging hours depending on charging current, battery capacity, and actual SOC of EV battery before charging. In recent years, a quick charging facility is also available that allows fast EV charging, ranging from 30 minutes to an hour [23]. This improves the flexibility of EV charging and increases the reliability for driving whenever required. However, longer plugged-in time in a charging station allows EV to deliver energy as a source and participate in providing ancillary services for a certain period of time.

Apart from V2D, there are various evolving notions of EV application in V2X which can be defined as follows:

- V2G allows feeding energy to the grid in the case of energy shortage in the grid, mainly during a lower power output from renewable energy sources (RESs).
- G2V allows obtaining energy from the grid and charging the vehicle. In addition to scheduled charging, EV can be charged during the peak generation periods of RESs.
- V2H allows meeting the partial or total load demand of a house. The preference can be given to peak periods only to maximize the economic benefit of EV for V2H.
- V2V allows EV to transfer its energy (charge/discharge) to another EV through the local grid or EV aggregator. Nonetheless, using the local grid may not be energy efficient considering the distance, time of charging, and costs.
- V2B performs similar to V2V/V2G but is limited to within the building. This feature allows efficient charge/discharge management of EVs and the energy management planning for the building (smart building management system).

These concepts of EV provide the flexibility of energy planning for home, building, and the grid. Thus EV will play a key role in the future smart grid. Nevertheless, in addition to the above concepts, V2D is the foremost priority before planning of EV for V2X. Therefore, all the roles of EV including the basic V2D are preferred for analysis, and their values, controls, practical outcome, and usefulness are explored in this study.

3. The structures of EV operation modes

This section discusses the structures of EV in different operation modes and the supports it may provide during the operational periods.

3.1 Structure of V2H

Typically, EV is preferred to be charged in the car park at home during overnight. This provides the flexibility to charge the vehicle and drive whenever necessary. With a bidirectional battery charger, the setup of EV for V2H can be formed. **Figure 1** shows the EV structure in V2H for exchanging energy with home energy demand. The same connection with the power conversion system can be used to charge EV (G2V) according to the specified energy management scheme. The power conversion system (PCS) allows feeding the energy demand of home partially or totally. PCS can be incorporated with small-scale RESs such as small wind turbine or roof-top solar generation panels through the central home controller. In **Figure 1**, the dashed blue line denotes EV discharging power flow for feeding home load demand (V2H). The solid blue line represents power flow direction to the grid (V2G), and solid green line symbolizes EV charging power flow (G2V). PCS provides the flexibility to regulate power flow and control charge/discharge of EV in G2V and V2H. Some of the very significant and unique features for V2H operation are as follows:

- V2H can involve one or two EVs and therefore comprises of significant amount of energy capacity to meet a typical home load demand.
- V2H facility is simple to implement, and some car manufacturers are already giving the opportunity to employ EV for V2H.

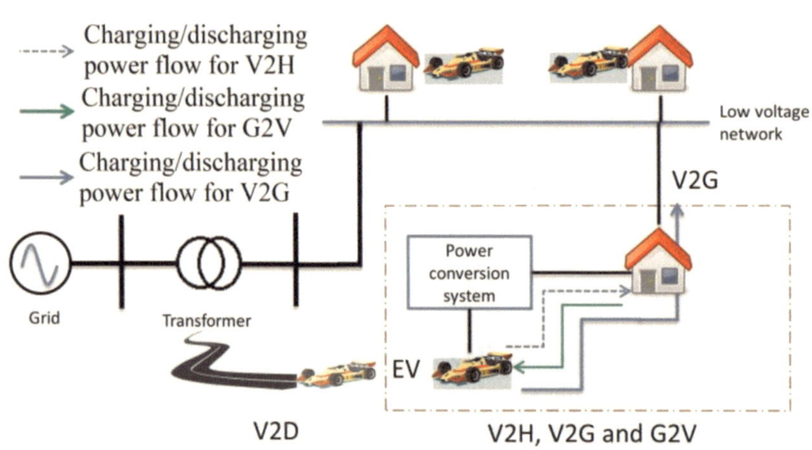

Figure 1.
Structure of V2D, V2H, V2G, and G2V at home.

- V2H can reduce peak demand and smooth home load demand profile or possibly can result in net zero energy purchase from the grid.
- Energy pricing based V2H can acquire financial gain to the owner and maximize the techno-economic benefit of EV for V2H.
- V2H is able to reduce the negative impact of RESs by operating as a storage device.
- The energy losses can be minimized through the use of V2H, thus increasing the efficiency of the grid.
- V2H can increase the operational flexibility in a smart home management system, thus improving the overall reliability of power supply at home.

3.2 Structure of V2G

A single EV with limited kWh capacity has an insignificant impact on the grid. However, as the number of EVs increases, a potential impact on grid level performance can be significant. The capacity could range from MWh to GWh, considering global target toward replacing the conventional fossil fuel-based cars with EVs. Hence, V2G structure comprises of a significant number of EVs connected to the grid. The structure of V2G is shown in **Figure 2**. The solid blue line and solid green line represent the power flow in V2G and G2V, respectively. EV may participate in V2G through an aggregator which comprises other necessary energy management

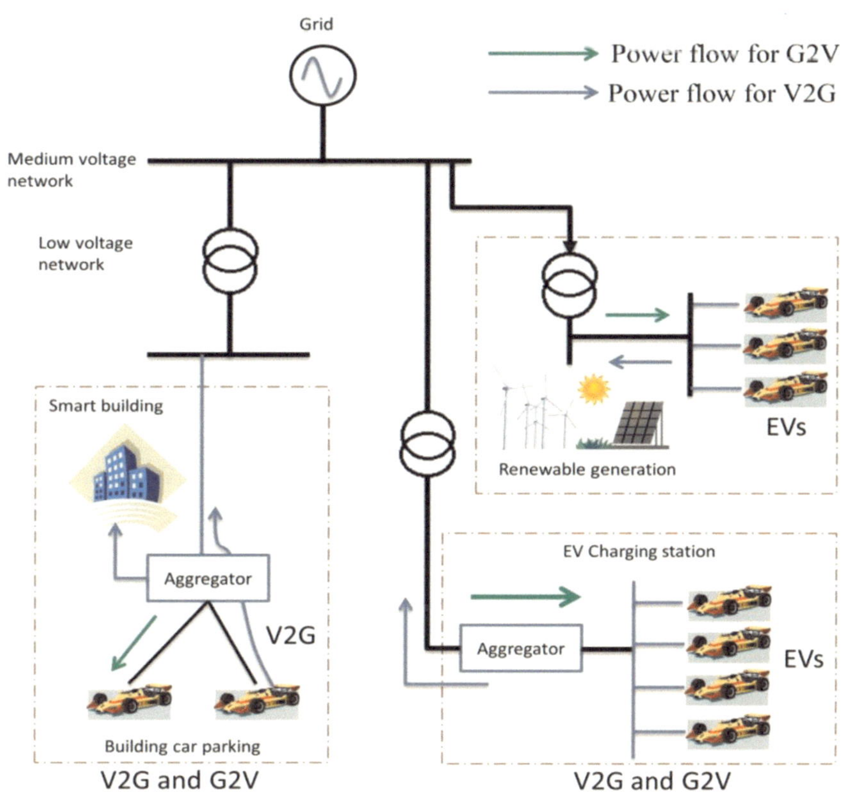

Figure 2.
Structure of V2G and G2V at commercial locations.

planning and controllers to facilitate V2G services. The aggregator can be placed at different locations such as charging station, parking lots, renewable energy farm, and smart building with other grid infrastructures to provide V2G services. Since V2G comprises many EVs, aggregator plays a key role in allocating and controlling power flows between various EVs. EVs for G2V can be connected to different grid voltage level through the associated transformers and distribution feeders. Mainly, EVs in a charging station are connected to the medium voltage network, whereas at home or buildings, EVs are connected to a low-voltage network. Some of the very significant and unique features for V2H operation are as follows:

- V2G comprises of a combination of many EVs, from a few to a hundred and more.
- The control of power flow of individual EVs is more complex but able to offer greater flexibility in energy regulation.
- V2G can provide both active and reactive power based on requirements.
- V2G through EV aggregator can provide a significant amount of power regulation capability to the grid.
- The large-scale EV battery storage provides the flexibility for energy planning based on RESs prediction and also in mitigating real-time power imbalances due to prediction error.
- In an isolated microgrid (MG), V2G can provide better stability and improved reliability performance, thus minimizing stress on baseload units.

3.3 Structure of V2B/V2V

V2B/V2V is of a similar kind of service as in V2G but in a reduced scale. V2V can be formed within a small community or a small isolated system to share energy according to suitable energy planning and available energy capacity as shown in **Figure 3**. V2V can add further flexibility in reducing grid variations due to integrated RESs in the community or MG. V2V can be implemented using direct V2V connection [24] or traditionally through the grid. On the other hand, V2B service is limited to a smart building, where EV can participate in overall energy management of the building. This allows integrating RESs in the building roof, regulating energy consumptions based on actual electricity pricing and maintaining owner-defined EV battery SOC. A greater perspective of energy planning, saving, economic gain, and EV SOC for V2D is possible to obtain through V2B. When EVs are operating in V2B/V2V, they can easily participate in V2G operation. The unique features of V2V/V2B can be summarized as follows:

- V2B/V2V consists of interaction between two or more EVs, and the control of EVs can be more complex than V2H, but it is simpler than V2G.
- V2B/V2V involves smart homes and car parking lots in a community or in a building for sharing energy among them.
- The energy losses in V2B/V2V vary according to individual vehicle location and the amount of energy sharing. The losses can be higher than V2H but they are lower than V2G.

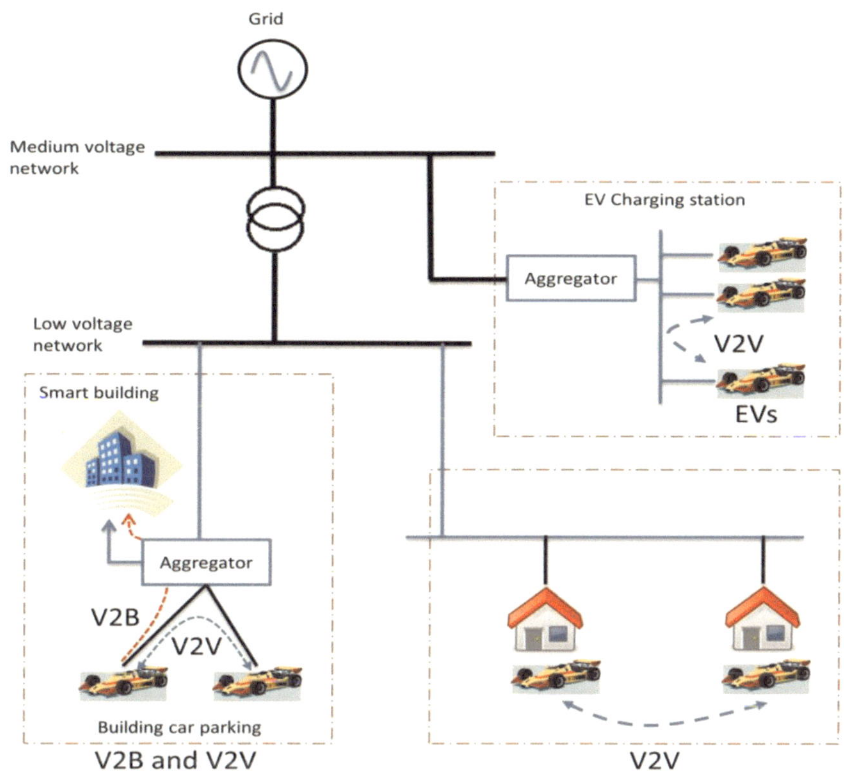

Figure 3.
Structure of V2V, V2B, and G2V at commercial locations.

- V2B/V2V can be coordinated with the RESs installed in the building or community houses and provide more reliable and secure energy supply for the consumer and in the end reduce stress to the grid.

- Since V2B/V2V is less complex but has access to large energy capacity, it can contribute significantly for implementing a smart grid.

4. Feedback regulation for V2X and ancillary services

The controller input for V2X varies according to particular ancillary services to the grid or home. The energy services can be ranged from under-frequency, over-frequency, and energy supplies to power smoothing provisions. However, EV battery SOC is the only parameter that defines the amount of energy capacity available for ancillary service regardless of the types of services. In V2H, mainly active power reference is used as feedback, whereas in V2B/V2G the reference could be extended to frequency, voltage, and other important parameters, as required by the grid operator.

4.1 Frequency-controlled EV regulation

Frequency regulation along with changing the load-generation situation with contingencies is crucial to control and maintain frequency balance and satisfy the grid requirements. A frequency feedback command to V2G controller regulates EV power output to respond to frequency variations for primary frequency control while maintaining EV owner preferred SOC limit [25]. The method can be used to regulate grid frequency by coordinating with the variation in RESs power output and minimize frequency oscillation [26]. In comparison to spinning reserves, frequency regulation is more frequent and is required for a short of time. A typical schematic of frequency feedback-controlled EV power regulation is shown in **Figure 4**. EV power output has regulated the deviation in frequency from the nominal value. The power-frequency (P-f) droop defines the specific EV power for the definite frequency deviation, and the droop gain adjusts the overall intensity of EV response. The set charging power of EV is denoted by P_{EV-set}^{Charge}. EV charging/discharging limit ensures that EV does not violate the maximum converter capacity. The output power reference at EV terminal (PEV/R) is executable when the defined SOC condition is satisfied, i.e., PEV/O has the absolute value of 1.

4.2 Voltage-controlled EV charging scheduling

With the increased roof-top solar penetration, grid voltage experiences spike especially during peak solar generation. Voltage-constrained EV charging plan can regulate the grid voltage considering variation in a solar generation as shown in **Figure 5** [27]. An energy pricing arrangement for voltage control in a fast charging

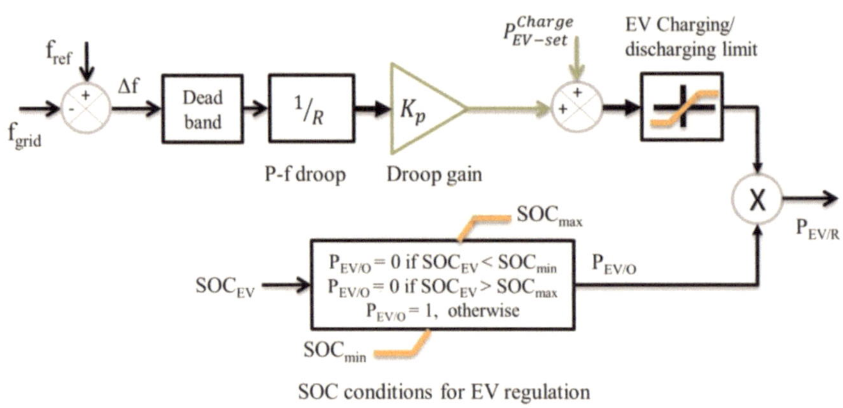

Figure 4.
Frequency-controlled EV regulation.

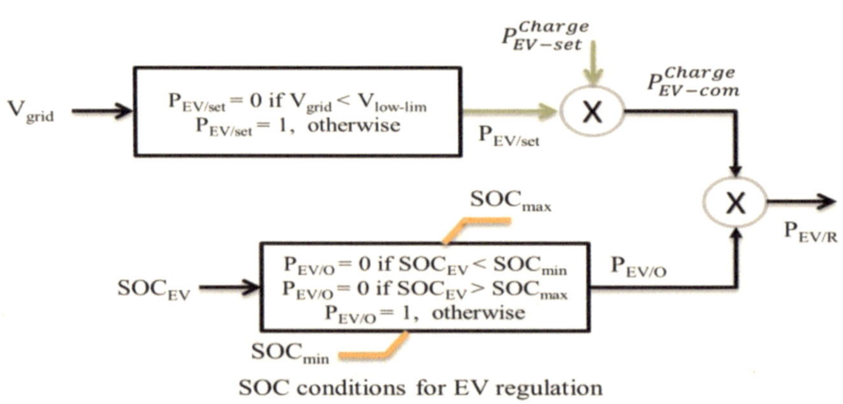

Figure 5.
Voltage-controlled EV charging scheduling.

station can also attract EV customer to charge the vehicle and participate in voltage control [28]. During peak solar generation in the daytime, EV charging will reduce the rising voltage distress from the peak generation by operating in charging mode. EV can be charged as long as the grid voltage is higher than the lower voltage limit, and hence the status of EV PEV/Set will be 1. EV charging command P_{EV-set}^{Charge} defines the power reference for EV charging to be executed, whereas PEV/O delineates the fulfillment of SOC conditions to maintain EV battery SOC during the charging/discharging process. EV can also be designed as V2G service provider (charge/discharge) during voltage transients as in frequency control using droop control method. EV can also participate in voltage feedback-controlled reactive power regulation of EV and enhance power quality issues in the grid and improve power quality of the grid by minimizing voltage sags with the discharge power from EV [29].

4.3 Power-controlled EV regulation

During peak shaving [30], in the case of load outage [31] or utilizing EV for mitigating power oscillation [32] resulted from RESs, the active power reference is taken as the feedback for V2G/V2H/V2B. The power output of EV is constrained by the converter capacity and SOC of EV battery. The active power output reference for EV can be written as in Eq. (1):

$$P_{ref-EV}(t) = P_{load\_e}(t) - P_{load\_a}(t) \tag{1}$$

where t is the time, $P_{ref\text{-}EV}(t)$ is the active power requirement at EV terminal to maintain power balance, $P_{load\_e}(t)$ is the estimated load demand, and $P_{load\_a}(t)$ is the actual load demand.

5. EV charge/discharge management (ECDM) and communication issues in smart grid

As mentioned in Section 4, the corresponding power equation of EV depends on the particular feedback signal. However, the calculation of SOC remains the same for any power equation of EV. The changes in battery energy during a charge/discharge process can be defined as in Eq. (2) [33]:

$$\Delta E_i = \int_0^t \eta P_t \, dt \tag{2}$$

Therefore, the actual SOC can be calculated from the SOC value at the previous stage and the percentage of the available energy capacity of EV battery in the following way as in Eq. (3):

$$SOC_t = SOC_{(t-1)} + \int_{t-1}^{t} \frac{\Delta E_i}{E_i} \, dt \tag{3}$$

where SOC_t is the SOC at time t, $SOC_{(t-1)}$ is the SOC at a time $(t-1)$, E_i is the battery capacity, η is the charging/discharging efficiency, and ΔE_i is the change in battery energy.

5.1 Modeling of ECDM in V2H and G2V

Since battery charger and battery are not 100% efficient, the amount of power available for discharging is less than the rated capacity, and the total intake power of EV during charging is higher than the rated capacity. Hence, the equation of EV power in V2H/G2V can be defined as in Eq. (4):

$$P_T(t) = P_D(t) + S(t) * \sum_{n=1}^{K} P_{EV,n}^{TC/TD}(t) \tag{4}$$

In V2H (discharge), the EV power output can be written as in Eq. (5):

$$P_{EV,n}^{TC/TD}(t) = -P_{EV,n}^{TD}(t) \tag{5}$$

In G2V (charge), the equation of EV power can be written as in Eq. (6):

$$P_{EV,n}^{TC/TD}(t) = P_{EV,n}^{TC}(t) \tag{6}$$

where $P_{EV,n}^{TC}(t) = \eta^C P_{EV,n}(t)$= total EV power in charging mode; $P_{EV,n}^{TD}(t) = \eta^D P_{EV,n}(t)$= total EV power in discharging mode; N = number of EVs connected to the home network; P_T = total home power demand; P_D = home load power demand without EV; S = EV position {at home and plugged-in (S = 1), outside/at home but unplugged (S = 0)}; $P_{EV,n}(t)$ = the power output of EV (positive (charging)/negative (discharging)); η^C = charging efficiency of EV charger and EV battery; η^D = discharging efficiency of EV charger and EV battery.

ECDM can be planned according to real-time pricing or randomly. However, unplanned charging will increase electricity costs of a home, which is the last thing an EV owner would wish for. Hence, an optimal charge/discharge plan needs to be adopted to minimize the cost of electricity in the home. The power calculation in V2H and G2V can be written as in Eq. (7):

$$P_{EV,n}^{TC/TD}(t) = \begin{cases} -P_{EV,n}^{TD}(t), & \begin{array}{c} \mathit{if}\ t = time_{on-peak}\ \mathit{and}\ P_{E\_on-peak}(t) > P_{E\_off-peak}(t) \\ \text{and } SOC(t) \geq SOC_{ODT} \text{ or } SOC(t) \geq SOC_{\min} \end{array} \\ P_{EV,n}^{TC}(t), & \begin{array}{c} \mathit{if}\ t = time_{off-peak}\ \mathit{and/or}\ P_{E\_off-peak}(t) < P_{E\_on-peak}(t) \\ \text{and } SOC(t) \leq SOC_{max} \end{array} \\ 0, & \mathit{otherwise} \end{cases} \tag{7}$$

where $P_{E\_on-peak}$= on-peak electricity price (per kWh); $P_{E\_off-peak}$ = off-peak electricity price (per kWh); SOC_{ODT} = EV owner defined SOC threshold (pu).

In the case of discharging EV during peak periods when electricity price is high and charging for the period of off-peak time with a lower price, the energy calculation can be represented as in Eq. (8):

$$\begin{aligned} \sum_{n=1}^{N} \sum_{0}^{t} P_{ET}(t) = & \sum_{off-peak_{min}}^{off-peak_{max}} \{P_D(t) + S(t) * P_{EV,n}^{TC}(t)\} * P_{E\_off-peak}(t) \\ & + \sum_{peak_{min}}^{peak_{max}} \{P_D(t) - S(t) * P_{EV,n}^{TD}(t)\} * P_{E\_on-peak}(t) \end{aligned} \tag{8}$$

where $P_{ET}(t)$= Total electricity price.

Without V2H (G2V only), total EV power in discharging mode is zero in Eq. (7), and hence the associated power reduction in Eq. (8) is zero.

5.2 Modeling of ECDM in V2G and G2V

The power equations of EV for V2G vary according to their installed location and the services it is providing. At home, EV must meet total home load demand, and then it can provide power to the grid. This is particularly applicable when a single connection point at home is available for the incoming and outgoing power. In V2G at home, the power equation can be written as in Eq. (9):

$$P_T(t) = P_D(t) - S(t) * \sum_{n=1}^{K} P_{EV,n}^{TD}(t) \tag{9}$$

where $\sum_{n=1}^{K} P_{EV,n}^{TD}(t) > P_D(t)$ and $P_T(t)$ is negative which indicates exporting power to the grid. The battery SOC conditions must be satisfied to secure customer preference and battery protection as in Eq. (10):

$$SOC(t) \geq SOC_{ODT} \text{ or } SOC(t) \geq SOC_{\min} \tag{10}$$

In view of selling energy to the grid when the price is high and recharging at off-peak periods with lower energy price, the calculation of energy pricing for V2G can be defined as in Eq. (11):

$$\begin{aligned} \sum_{n=1}^{N} \sum_{0}^{t} P_{ET}(t) = & \sum_{off-peak_{min}}^{off-peak_{max}} \left\{ P_D(t) + S(t) * P_{EV,n}^{TC}(t) \right\} * P_{E\_off-peak}(t) \\ & + \sum_{peak_{min}}^{peak_{\max}} \left\{ P_D(t) - S(t) * P_{EV,n}^{TD}(t) \right\} * P_{SE\_on-peak}(t) \end{aligned} \tag{11}$$

where $P_{SE\_on-peak}$ = on-peak electricity selling price (per kWh).

At charging stations, V2G can be implemented as a power reference in the feedback loop of EV power regulation or frequency regulation through droop/inertia. In this case, the calculation of EV power can be defined as in Eq. (12):

$$P_{EV,n}^{TC/TD}(t) = \begin{cases} -P_{EV,n}^{TD}(t), & V2G \\ P_{EV,n}^{TC}(t), & G2V \\ 0, & otherwise \end{cases} \tag{12}$$

In the case of frequency regulation in the feedback loop, the droop-controlled equation can be defined as in Eq. (13):

$$P_{EV,n}^{TC/TD}(t) = \begin{cases} -P_{EV,f}^{TD}(t), & \mathrm{f}_{actual} < \mathrm{f}_{ref} \\ P_{EV,f}^{TC}(t), & \mathrm{f}_{actual} > \mathrm{f}_{ref} \ or \ SOC(t) < SOC_{\max} \\ P_{EV,r}^{TC}(t), & SOC(t) < SOC_{\min} \ or \ SOC(t) < SOC_{RTD} \ or \ SOC_{\max} \\ 0, & SOC(t) > SOC_{\min} \ or \ SOC_{ODT} \ or \ SOC_{\max} \end{cases} \tag{13}$$

where

$$P_{EV,f}^{TD}(t) = P_{EV,f}^{TC}(t) = \frac{\mathrm{f}_{actual} - \mathrm{f}_{ref}}{R} = \frac{\Delta f}{R}$$

$P^{TD}_{EV,f}(t)$ = EV discharge power in V2G for a frequency lower than the rated value; $P^{TC}_{EV,f}(t)$ = EV charge power in V2G for a frequency higher than the rated value; $P^{TC}_{EV,r}(t)$ = the rated EV charging power in (a) V2G when EV is inactive (frequency is within the allowed limit) in V2G priority mode or (b) in the recharging priority mode, when SOC is lower than the minimum SOC/required SOC threshold for driving/maximum SOC; Δf = frequency deviation; SOC_{RTD} = required SOC threshold for driving back to home from the workplace.

Hence, the available SOC for V2G service is as in Eq. (14):

$$SOC_{V2G} = SOC_{RTD} - SOC(t) \tag{14}$$

The inertia and droop control can be combined to obtain improved frequency response [34]. Thus, the power reference to reflect charge/discharge in frequency regulation can be expressed as in Eq. (15):

$$P^{TD}_{EV,f}(t) = P^{TC}_{EV,f}(t) = \frac{f_{actual} - f_{ref}}{R} + \frac{d}{dt}\Delta f\, G_{in} = \frac{\Delta f}{R} + \frac{d}{dt}\Delta f\, G_{in} \tag{15}$$

where G_{in} is the gain of the inertial controller.

5.3 Modeling of ECDM in V2B/V2V

The feedback loop for EV in V2B/V2V modeling can be the same as in V2G. The modeling depends on the purpose and control in building energy management system for V2B or simple energy exchange between two or more EVs. Hence, power regulation as in Eq. (12) or frequency regulation as in Eq. (13) can be picked out for V2B. In V2V, simple power regulation for exchanging energy can be selected as in Eq. (12). The exchanged energy of an EV in V2G at charging stations/building (V2B)/other vehicle (V2B) can be easily priced according to Eq. (16):

$$\sum_{0}^{t} P_{ET}(t) = P^{TC}_{EV}(t) * P_{charge}(t) - P^{TD}_{EV}(t) * P_{discharge}(t) \tag{16}$$

where P_{charge} is the energy price for charging (G2V) and $P_{discharge}$ is the energy price for discharging (V2G). The negative and positive prices indicate credit and debt, respectively, for EV owner.

5.4 Modeling of ECDM in V2D

The energy consumption in V2D depends on certain aspects related to the vehicle as well as weather and road conditions. The tractive effort (F_{te}) to propel the vehicle forward must overcome and accomplish the forces as in Eq. (17) [35]:

$$F_{te} = F_a + F_{rr} + F_{ad} + F_{gr} \tag{17}$$

where F_a = acceleration force, F_{rr} = rolling resistance force, F_{ad} = aerodynamic force, and F_{gr} = grade resistance force.

The acceleration force of EV can be derived from Newton's law as in Eq. (18):

$$F_a = ma \tag{18}$$

where m and a are vehicle mass (kg) and acceleration (m/s^2), respectively.

The rolling resistance is related to the frictions related to the vehicle tire on road, bearings, and gearing system. This is roughly invariable with vehicle speed and proportional to vehicle mass that can be derived as in Eq. (19):

$$F_{rr} = maC_{rr} \tag{19}$$

where C_{rr} is the coefficient of rolling resistance.

The aerodynamic force which is a function of the design of the car that represents the friction of EV in the air can be formulated as in (Eq. (20)):

$$F_{ad} = 0.5\rho A C_d V^2 \tag{20}$$

where ρ is the surrounding air density (kg/m^3); A is the front area of the car (m^2); C_d is the drag coefficient that depends on the design of the car, shape, and frontal area of the car; and V is the velocity.

The grade resistance force can be defined as vehicle weight that functions along the slope as in Eq. (21):

$$F_{gr} = ma \sin \phi \tag{21}$$

where ϕ is the slope angle.

Thus, the mechanical power of EV to overcome the aforementioned forces can be calculated as in Eq. (22):

$$P_{tm} = maV + maV \sin \phi + 0.5\rho A C_d V^3 + maC_{rr}V \tag{22}$$

5.5 EV SOC control in various operation modes

Regardless of different operation modes, EV SOC changes as it consumes or provides energy and therefore needs to be recharged for guaranteeing the availability of EV for the next service. As EV charger and battery are not 100% energy efficient, the total energy capacity of EV for V2X is not available and can be calculated according to the efficiency of battery and EV charger. Also, energy consumed by EV including the charger is slightly higher than the rated capacity. While V2B, V2G, and V2V are independent of distance, V2D is solely dependent on traveling distance and speed, in general.

As outlined in Section 5.4, EV SOC in V2D is subjected to certain variables that are specific to the vehicle such as vehicle mass, rolling resistance, vehicle aerodynamics, specific road conditions such as down/up slop, the number and behavior of passengers and the driver, air conditioner/heater on/off, and day/night time (headlight on/off). In addition to these, vehicle speed defines the change in energy capacity. In Australia, the average traveling distance is less than 20 km for 73% of the employed people traveling from their working place to their residing place [36]. Moreover, passenger vehicles in Victoria annually traveled 14,498 km on average [37]. Thus, an average daily distance by individual passenger vehicle is approximately 55 km in view of 260 working days. Therefore, the SOC calculation based on distance can be written as in Eq. (23):

$$SOC_{DT} = SOC_{(DT-1)} + \frac{\Delta E_{DT}}{E_i} \tag{23}$$

where SOC_{DT} is the SOC at present location (km), $SOC_{(DT-1)}$ is the SOC at previous location (km), and ΔE_{DT} is the amount of energy consumed for the traveled distance per km. In V2G/G2V/V2B/V2H, battery SOC is calculated by Eq. (3).

5.6 Communication issues of EV in smart grid

In a smart city environment, where centralized or dispersed energy generations and load systems are connected to the web through wireless or wired connections, communication issues and standards play an important role for the secure and reliable exchange of data and information between various entities. Therefore, scalable and efficient communication technologies are imperative for EV to be a part of the intelligent transportation system and electric grid [38]. There are multiple communication standards that are in use in existing studies such as ISO 15118 [39], IEEE 1609 WAVE, and IEC 61850 [40]. Nevertheless, the performance evaluation studies in [40, 41] demonstrated the superiority of IEC 61850 standards over the other. Hence, it can be said that standard communication technologies are the key to implement smart grid facility and attain maximum energy benefit of the integrated energy elements.

6. Case studies of EV in various operation modes

This section presents various cases of EV charge/discharge management in providing multiple services.

6.1 EV SOC in V2D, G2V, and V2H at home

When EV participates in different operation modes, the feedback control strategy as mentioned in Section 4 defines EV power regulation. In addition to the feedback control strategy, an energy pricing policy can also be included in the control approach for economic analysis and ensuring financial benefit for the EV owner for EV battery charging/discharging. The power exchange of an EV for a typical day in different EV operation modes is presented in **Figure 6**. EV discharging efficiency for V2X is considered as 98%. It is assumed that EV owner travels a distance of 55 km/day as mentioned in Section 5.5. The selected EV model is Nissan LEAF, and discharged power during V2D is collected from the reference [42]. EV needs to be charged to the maximum SOC (1.0pu) before the next morning, and considering economic aspect, EV is charged during off-peak periods only. The consuming and supplying of power are denoted by positive and negative values, respectively. The daily traveling time of the vehicle is 8–9 am in the morning and 5–6 pm in the afternoon. The vehicle speed is 80 km/h for 20 km, 90 km/h for 15 km, and 60 km/h for 20 km, and the associated EV power consumption is 152, 163, and 128 Wh/km, respectively [33]. The EV discharges 8.045 kWh of energy for

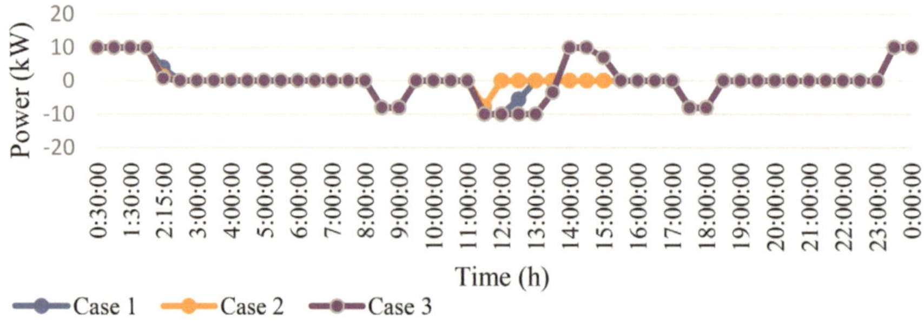

Figure 6.
The power exchange of an EV for a typical day in various operation modes.

one-way travel, and thus battery SOC reduces to 0.799 pu when EV arrives at the work place. The EV SOC reduces by 0.201 pu in the morning, and assuming similar distance and driving speed, the required battery energy capacity is the same to return back home. Hence, with the minimum SOC of 0.2 pu for safe battery discharge limit, the updated SOC value before the departure from office has to be greater or equal to 0.401 pu. It is assumed that V2G operation starts at 11 am and continues for multiple hours in accordance with the selected SOC constraints. There are three SOC constraints in the study, and they are as follows: (1) V2G (with V2V and V2B) and G2V, Case 1; (2) V2G (with V2V and V2B), V2H, and G2V, Case ; and (3) V2G (with V2V and V2B), V2H, and G2V (recharge), Case 3. EV is fully charged during overnight when the energy price is low. EV battery SOC status and constraints for three cases are outlined in **Table 1**.

Case 1 indicates that when EV is not participating in V2H, higher amount of energy is available for V2G/V2V/V2B services compared to Case 2 as shown in **Figure 6** and **Table 1**. However, this will drain out all the EV energy, i.e., EV battery SOC reaches to the lowest safe margin when EV arrives at home as shown in **Figure 7** and may not be accessible in the case of emergency driving. The less amount of energy is obtainable for V2G/V2V/V2B services in the event of Case 2 as EV is planned to be utilized for V2H. Therefore, EV battery SOC is higher than Case 1. This has multiple benefits, i.e., EV can be placed for V2H or can be accessible for emergency V2D in the evening. Nevertheless, available energy for V2G/V2V/V2B services reduces by nearly half than that of Case 1. EV SOC reduces gradually as EV discharges in V2H operation. Considering such contrasting environment of Case 1 and Case 2, a new charge/discharge plan is presented, i.e., Case 3. In the instance of Case 3, EV is discharged until battery SOC reaches to the minimum SOC level as shown in **Figure 7** and then recharged back to the expected value of 0.601pu before

Case	SOC on arrival at office	Plan for V2H and estimated SOC (pu) reserve for V2H	Expected SOC (pu) on arrival at home	SOC (pu) for traveling back to home	SOC (pu) before departure with minimum SOC 0.2pu	Energy (kWh) available for V2G/V2V/V2B
Case 1	0.799	No	0.2	0.201	0.401	15.592
Case 2	0.799	Yes/0.2	0.4	0.201	0.601	7.752
Case 3	0.799	Yes/0.2	0.4	0.201	0.601	23.476

Table 1.
EV SOC status and constraints at various operation modes.

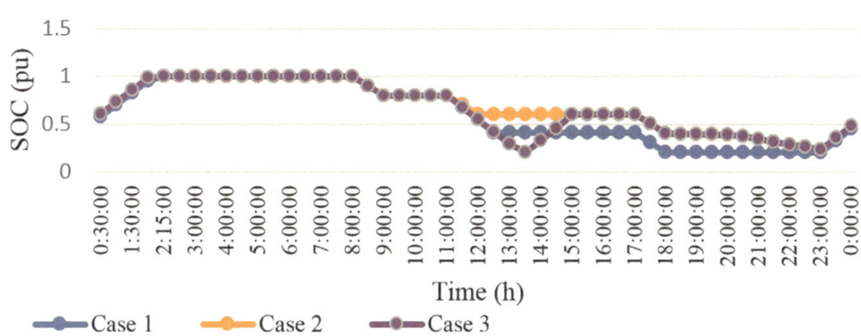

Figure 7.
EV SOC status for a typical day in various operation modes.

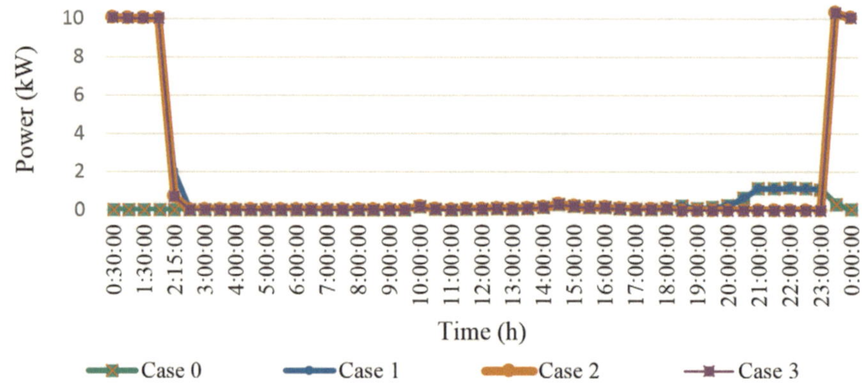

Figure 8.
EV power in V2H and G2V at home only.

the departure of EV in the afternoon. This ensures that higher amount of energy (23.476kWh) is available for V2G/V2V/V2B services, and also sufficient SOC is available when EV arrives at home for V2H or emergency V2D.

The amount of energy exchange at home in different EV operation modes is shown in **Figure 8** where Case 0 defines the event of energy exchange without an EV. For Cases 1, 2, and 3, the maximum charging power is 10 kW during off-peak time to reduce EV charging cost. The close-up view shown in **Figure 9** illustrates that total house energy demand between 18:00 and 23:00 h is met by EV, and hence the net cost of energy purchase from the grid is zero.

Hence, it is observed that EV can participate in various operation modes according to the planning and management of EV battery SOC. An effective SOC management such as Case 3 provides better resolution on battery SOC management and maximum utilization of EV battery capacity in a smart grid environment.

6.2 Aggregated EV for V2G services

In order to evaluate the performance and importance of EV in the power system for V2X services, an isolated MG is considered as shown in **Figure 9** with different level of EV penetration. A 5.1 MW PV farm is integrated at bus 3, and EVs are connected through an aggregator at bus 2. Two different case studies are simulated to demonstrate the contribution of EV in V2G. A 100% load growth is applied at bus 3 for the duration of 0–0.5 s for various scenarios, and the generator frequency responses are shown in **Figure 10**. **Figure 10** depicts that with integrated PV (dotted green line), the frequency of generator drops to 0.9874pu in comparison to 0.9925pu in the case of without any PV integration (solid blue line). The frequency

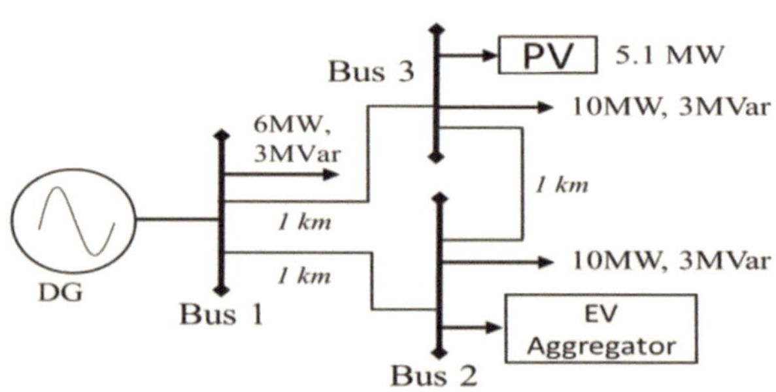

Figure 9.
Studied MG with PV and aggregated EV.

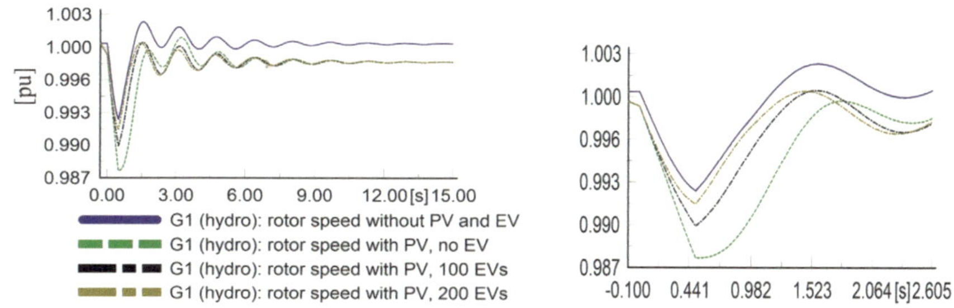

Figure 10.
The generator frequency response with 100% load growth at bus 3.

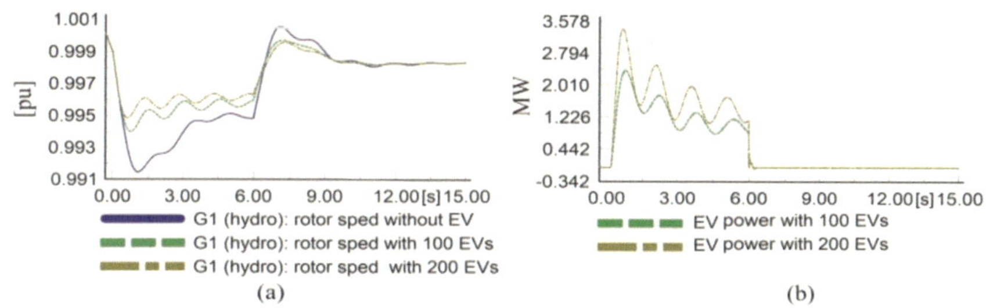

Figure 11.
The generator frequency response with temporary PV outage.

drops due to the fact of negative inertial contribution with increased PV penetration. On the contrary, the frequency drop reduces considerably with EV participation in V2G regulation. It can also be seen that with the increased number of EVs (100 vs. 200), frequency deviation reduces as EV aggregator provides more power with a higher amount of EVs. This implies better frequency regulation and energy management in a smart grid environment.

Another case study on the temporary outage of PV power output during t = 0–2 s due to cloudy weather condition is considered to further exhibit EV importance in V2G regulation. **Figure 11a** also illustrates an improved frequency performance, i. e., lower frequency deviation with the integrated EV (dotted green and dark yellow lines) than without an EV (solid blue line) while providing V2G services. The higher amount of power output of EV aggregator is visible in **Figure 11b** as the number of EV increases which justifies the improved frequency performance following the disturbance events.

7. Conclusion

With revolutionized transportation industry, numbers of EVs are expected to rise around the globe. However, the large volumes of EVs are a great concern for stable and reliable operation of grid due to the unpredictable moving nature of EV. Hence, it is essential to adopt suitable planning for the management of EV energy. V2X is one of the paramount means of improving power systems' stability and reliability, power quality, and maximizing the economic benefit of EV in the present-day and future electric grid. This chapter has thoroughly presented and discussed the structures of EV in various operation modes, feedback regulation, and the charge/discharge management strategy of EV for V2X services. Furthermore, a case study demonstrates the flexibility of EV charge/discharge planning and management at grid level and customer end in accordance with associated EV battery

SOC status and constraints. In addition, this study illustrates the ample opportunity of EV in V2X and G2V to minimize the technical challenges from the expected enormous penetration of EV in the coming years.

Acknowledgements

This work is supported by Victoria University International Post-graduate Research Scholarship scheme.

Nomenclature

GHG	greenhouse gas
EV	electric vehicles
V2H	vehicle to home
V2G	vehicle to grid
V2V	vehicle to vehicle
V2B	vehicle to building
V2D	vehicle to drive
G2V	grid to vehicle
SOC	state of charge
MG	microgrid
ECDM	EV charge/discharge management
PCS	power conversion system
V2X	vehicle to everything

Author details

Ujjwal Datta*, Akhtar Kalam and Juan Shi
College of Engineering and Science, Victoria University, Melbourne, Australia

*Address all correspondence to: ujjwal.datta@live.vu.edu.au

References

[1] Fuquan Z, Feiq I, Zongwei L, Han H. The correlated impacts of fuel consumption improvements and vehicle electrification on vehicle greenhouse gas emissions in China. Journal of Cleaner Production. 2019;**207**:702-716

[2] Chiu CO, Nuruol SM, Choon WY, Siaw CL, Suhana K, Ahmad FAR, et al. Greenhouse gas emissions associated with electric vehicle charging: The impact of electricity generation mix in a developing country. Transportation Research Part D: Transport and Environment. 2018;**64**:15-22

[3] Youngmin K, Byung-In K, Young MK, Hyemoon J, Jeongin K. Charging scheduling problem of an M-to-N electric vehicle charger. Applied Mathematical Modelling. 2018;**64**: 603-614

[4] Liu T, Zou Y, Liu D, Sun F. Reinforcement learning of adaptive energy management with transition probability for a hybrid electric tracked vehicle. IEEE Transactions on Industrial Electronics. 2015;**62**:7837-7846

[5] Shi C, Tang Y, Khaligh A. A single-phase integrated onboard battery charger using propulsion system for plug-in electric vehicles. IEEE Transactions on Vehicular Technology. 2017;**66**:10899-10910

[6] Wang N, Tang L, Pan H. A global comparison and assessment of incentive policy on electric vehicle promotion. Sustainable Cities and Society. 2019;**44**: 597-603

[7] Wang Q, Liu X, Du J, Kong F. Smart charging for electric vehicles: A survey from the algorithmic perspective. IEEE Communications Surveys & Tutorials. 2016;**18**:1500-1517

[8] Saldanha JJA, Santos EMD, Mello APCD, Bernardon DP. Control strategies for smart charging and discharging of plug-in electric vehicles. In: Silva IND, editor. Smart Cities Technologies. London: IntechOpen; 2016. pp. 121-141

[9] Nie Y, Wang X, Cheng KE. Multi-area self-adaptive pricing control in smart city with EV user participation. IEEE Transactions on Intelligent Transportation Systems. 2018;**19**: 2156-2164

[10] Calvillo CF, Sánchez AM, Villar J, Martín F. Impact of EV penetration in the interconnected urban environment of a smart city. Energy. 2017;**141**: 2218-2233

[11] Shuai W, Maillé P, Pelov A. Charging electric vehicles in the smart city: A survey of economy-driven approaches. IEEE Transactions on Intelligent Transportation Systems. 2016;**17**:2089-2106

[12] Berthold F, Ravey A, Blunier B, Bouquain D, Williamson S, Miraoui A. Design and development of a smart control strategy for plug-in hybrid vehicles including vehicle-to-home functionality. IEEE Transactions on Transportation Electrification. 2015;**1**: 168-177

[13] Xu NZ, Chung CY. Reliability evaluation of distribution systems including vehicle-to-home and vehicle-to-grid. IEEE Transactions on Power Systems. 2016;**1**:31-41

[14] Tushar MHK, Assi C, Maier M, Uddin MF. Smart microgrids: Optimal joint scheduling for electric vehicles and home appliances. IEEE Transactions on Smart Grid. 2014;**5**:239-250

[15] Xiaohua W, Xiaosong H, Scott M, Xiaofeng Y, Volker P. Stochastic control of smart home energy management with plug-in electric vehicle battery energy

storage and photovoltaic array. Journal of Power Sources. 2016;**333**:203-121

[16] Colmenar SA, Rodriguez PC, Enrique AR, Borge DD. Estimating the benefits of vehicle-to-home in islands: The case of the Canary Islands. Energy. 2017;**134**:311-322

[17] Karfopoulos EL, Hatziargyriou ND. Distributed coordination of electric vehicles providing V2G services. IEEE Transactions on Power Systems. 2016; **31**:329-338

[18] Monteiro V, Exposto B, Ferreira JC, Afonso JL. Improved vehicle-to-home (iV2H) operation mode: Experimental analysis of the electric vehicle as off-line UPS. IEEE Transactions on Smart Grid. 2017;**8**:2702-2711

[19] Melo HND, Trovão JPF, Pereirinha PG, Jorge HM, Antunes CH. A controllable bidirectional battery charger for electric vehicles with vehicle-to-grid capability. IEEE Transactions on Vehicular Technology. 2018;**67**:114-123

[20] Datta U, Saiprasd N, Kalam A, Shi J, Zayegh A. A price regulated electric vehicle charge-discharge strategy. International Journal of Energy Research. 2018;**43**:1032-1042

[21] Inside EVs. 7 Electric Cars With The Biggest Batteries. 2018. Available from: https://insideevs.com/seven-electric-cars-biggest-batteries/ [Accessed: December 10, 2018]

[22] Gorzelany J. Here Are The 10 Longest Range Electric Cars. 2018. Available from: https://insideevs.com/10-longest-range-electric-cars-available-in-the-u-s/ [Accessed: December 10, 2018]

[23] Motors M. i-MiEV. 2018. Available from: https://www.mitsubishi-motors.com/en/showroom/i-miev/catalog/pdf/17_5my_i_miev_g_exp.pdf. [Accessed: December 10, 2018]

[24] Masrur MA, Skowronska AG, Hancock J, Kolhoff SW, McGrew DZ, Vandiver JC, et al. Military-based vehicle-to-grid and vehicle-to-vehicle microgrid—system architecture and implementation. IEEE Transactions on Transportation Electrification. 2018;**4**: 157-171

[25] Hernández JC, Sutil FS, Vidal PG, Casas CR. Primary frequency control and dynamic grid support for vehicle-to-grid in transmission systems. International Journal of Electrical Power & Energy Systems. 2018;**100**:152-166

[26] Liu H, Hu Z, Song Y, Lin J. Decentralized vehicle-to-grid control for primary frequency regulation considering charging demands. IEEE Transactions on Power Systems. 2013; **28**:3480-3489

[27] Cheng L, Chang Y, Huang R. Mitigating voltage problem in distribution system with distributed solar generation using electric vehicles. IEEE Transactions on Sustainable Energy. 2015;**6**:1475-1484

[28] Xiaohong D, Yunfei M, Xiandong X, Hongjie J, Jianzhong W, Xiaodan Y, et al. A charging pricing strategy of electric vehicle fast charging stations for the voltage control of electricity distribution networks. Applied Energy. 2018;**225**:857-868

[29] Brenna M, Foiadelli F, Longo M. The exploitation of vehicle-to-grid function for power quality improvement in a smart grid. IEEE Transactions on Intelligent Transportation Systems. 2014;**15**: 2169-2177

[30] Wang Z, Wang S. Grid power peak shaving and valley filling using vehicle-to-grid systems. IEEE Transactions on Power Delivery. 2013;**28**:1822-1829

[31] Shin H, Baldick R. Plug-in electric vehicle to home (V2H) operation under a grid outage. IEEE Transactions on Smart Grid. 2017;**8**:2032-2041

[32] Alam MJE, Muttaqi KM, Sutanto D. Effective utilization of available PEV battery capacity for mitigation of solar PV impact and grid support with integrated V2G functionality. IEEE Transactions on Smart Grid. 2016;**7**: 1562-1571

[33] Masuta T, Yokoyama A. Supplementary load frequency control by use of a number of both electric vehicles and heat pump water. IEEE Transactions on Smart Grid. 2012;**3**: 1253-1262

[34] Almeida PR, Soares F, Lopes JP. Electric vehicles contribution for frequency control with inertial emulation. Electric Power Systems Research. 2015;**127**:141-150

[35] Larminie J, Lowry J. Electric Vehicle Technology Explained. West Sussex: John Wiley & Sons Ltd; 2003

[36] Australian Bureau of Statistics. ABS Census of Population and Housing. 2016. Available from: http://www.abs.gov.au/ausstats/abs@.nsf/Lookup/by%20Subject/2071.0.55.001~2016~Main%20Features~Commuting%20Distance%20for%20Australia~1. [Accessed: December 14, 2018]

[37] Australian Bureau of Statistics. Survey of Motor Vehicle Use, Australia. 2016. Available from: http://www.abs.gov.au/ausstats/abs@.nsf/mf/9208.0. [Accessed: December 14, 2018]

[38] Wager G, Whale J. Driving electric vehicles at highway speeds: The effect of higher driving speeds on energy consumption and driving range for electric vehicles in Australia. Renewable and Sustainable Energy Reviews. 2016; **60**:158-165

[39] Cai L, Pan J, Zhao L, Shen X. Networked electric vehicles for green intelligent transportation. IEEE Communications Standards Magazine. 2017;**1**:77-83

[40] Schürmann D, Timpner J, Wolf L. Cooperative charging in residential areas. IEEE Transactions on Intelligent Transportation Systems. 2017;**18**: 834-846

[41] Hussain SMS, Ustun TS, Nsonga P, Ali I. IEEE 1609 WAVE and IEC 61850 standard communication based integrated EV charging management in smart grids. IEEE Transactions on Vehicular Technology. 2018;**67**: 7690-7697

[42] Ustun TS, Hussain SMS, Kikusato H. IEC 61850-based communication modeling of EV charge-discharge management for maximum PV generation. IEEE Access. 2019;**7**: 4219-4231

Chapter 9

Forecasting Recharging Demand to Integrate Electric Vehicle Fleets in Smart Grids

Juan Ignacio Guerrero Alonso, Enrique Personal, Antonio Parejo, Sebastián García, Antonio García and Carlos León

Abstract

Electric vehicle fleets and smart grids are two growing technologies. These technologies provided new possibilities to reduce pollution and increase energy efficiency. In this sense, electric vehicles are used as mobile loads in the power grid. A distributed charging prioritization methodology is proposed in this paper. The solution is based on the concept of virtual power plants and the usage of evolutionary computation algorithms. Additionally, the comparison of several evolutionary algorithms, genetic algorithm, genetic algorithm with evolution control, particle swarm optimization, and hybrid solution are shown in order to evaluate the proposed architecture. The proposed solution is presented to prevent the overload of the power grid.

Keywords: smart grids, vehicle-to-grid, electric vehicles, charging prioritization, electric vehicle fleets, evolutionary computation

1. Introduction

The electric vehicle (EV) means a new research field in smart grid (SG) ecosystems [1]. Currently, the new generation of EV provides different technologies that can be integrated in SGs [2]. However, these new technologies make the distribution grid management difficult [3, 4]. In particular, the EVs and the infrastructure needed to charge them have provided a great quantity of new standards and technologies. Currently, there are several research lines related to EVs, fast-charging networks (e.g., see [5], battery performance modeling [6], parasitic energy consumption, EV promotional policies, and increase in the range of the battery in EV [7], and other research lines related to EV energy management, contract models for consumption vehicle, market model to adopt the EVs, distributed energy resources management systems (DERMS), DER standards, faster charging technologies, demand response management systems (DRMS), the role of aggregators in the V2G (vehicle-to-grid), and energy efficiency (e.g., see [8], customer support, driver support, etc.). Additionally, all these lines are influenced by current regulations, and it could be very different among countries (e.g., the regulation between United States and Europe is very different in energy management). The charging infrastructure affects the SG on several levels. These levels concern transportation, distribution, and retailer levels. The main affected frameworks inside these levels are energy management, distribution management, and demand response. The energy management systems compound several functions [9]. One of them is the control of energy flows. The charging of EV can be made in any point of the grid that has a charging unit. If the system has information about the expected use of the charging unit, the energy flow will be easier to manage [10, 11]. The distribution management is related to distribution system operators (DSO). Usually, the charging infrastructure oversees DSOs. Thus, the DSOs must manage these facilities and maintain the information about them. Finally, the demand response concerns retailers and DSOs, and the main problem is demand curve flattening and price management [12]. Nevertheless, the new paradigm proposed by standard organizations, like National Institute of Standards and Technology (NIST), International Electrotechnical Commission (IEC), etc., related to the V2G proposed that the EV could charge or discharge the batteries [13]. Thus, the EV is a power source in specific scenarios. In these cases, the distributed resource management is affected by the new V2G technologies as a distributed power resource in low voltage without total availability, like some renewable energy resources, for example, wind and solar energy [14].

This paper proposed a solution for fleet charging prioritization, based on the concept of virtual power plant (VPP) and using distributed evolutionary computation algorithms to optimize the prioritization of EV fleets at different levels of SG ecosystems. A comparison of different evolutionary algorithms is performed.

This paper shows the proposed solution, starting with a bibliographical review. Then the architecture over different levels of SG is described, including the information flows. The evolutionary algorithm is described at different levels of SG ecosystem. Finally, the results of test the evolutionary algorithms are shown.

2. Bibliographic review

There are several research lines that are related with EVs, involving batteries (e.g., in [15], renewable energy [16], battery management systems [17, 18], energy management systems [19], charging spots [20, 21], driver assistance system [22], etc.). The EVs in the last millennium [23] provided a scenario with several vehicles with different types of vehicles: EVs, hybrid electric vehicles (HEVs), fuel cell vehicles, etc. The introduction of EVs provides several advantages, like reduction in greenhouse gas emissions [24]. But people's acceptance is necessary of EVs for daily usage [85, 86] and to have additional energy resources [23] in order to include the associated infrastructure. Additionally, the acceptance of EVs for daily usage, one of the first EVs, was the hybrid electric vehicle (HEV), and different types of HEVs were designed to reduce power requirements and increase vehicle autonomy, charging duration, and energy efficiency, selecting the appropriate battery [25]. Additionally, there are several studies about the performance of batteries in Ref. [26] and battery degradation in [27]. The new generation of EVs has several requirements not only in power but also infrastructure in Ref. [28]. There are several studies to establish renewable energy sources to support the charging of EVs and HEVs and cover the power requirements, for example, based on wind energy [87, 88], photovoltaic resources [29], general congestion of EV charging based on renewables [30], etc. However, SGs have provided a good scenario to integrate EVs and charging infrastructure.

Another solution could be the application of the queue theory [31]. The queue theory has an application in several topics: boarding management in [32], healthcare in Refs. [89, 90], dynamic facility layout problem in [33], optimization of traffic by means of signal-controlled management in [34], data acquisition in [35], etc. Usually, these references manage only one queue or several independent queues. However, there are more complex problems based on distributed software systems [36], which provide more difficult applications of queue theory. In these cases, queue theory should be adapted to a distributed environment. This paper proposes a novel solution to avoid this complexity.

There are other manners to manage the EV fleets that involve directly or indirectly the demand concept, improving the accuracy of energy forecasting [37]. The driver pattern modeling could improve fleet management, increasing the efficiency and sustainability and it could be used to forecast demand, some of these cases are: vehicular driving patterns in the Edinburgh region and to offer an option of battery electric vehicles for sustainable mobility is estimated in Ref. [38]; usage patterns and the user perception are the main objectives of a longitudinal assessment of the viability of EV for daily use study in Ref. [39]; the usage of autonomous vehicles and eco-routing like in [40]; or, ridesharing of shared autonomous vehicle fleet [41]. There are other examples oriented to transport management which includes in different ways the concept of demand: a bi-level optimization framework for EV fleet charging based on a realistic EV fleet model including a transport demand submodel is proposed in Ref. [42]. Other references treated the problem from the point of view of the congestion management of the electrical distribution network in case of limited overall capacity, for example, a distributed control algorithm for optimal charging is proposed in reference [43], or depending on the routing problem [44], allowing partial battery recharging with hybrid fleets (conventional and electric vehicles). Other references provided a solution to integrate renewable energy sources, investigating the possibilities to integrate additionally loads of uncertain renewable energy sources by smart-charging strategies as is proposed in Ref. [45].

There are several algorithms which provide solutions related to peak saving in demand curve. A real-time EV smart-charging method that not only considers currently connected EVs but also uses a prediction of the EVs that are expected to plug in the future is proposed in Ref. [46].

The authors in Ref. [47] propose VPPs as a new solution for the implementation of technologies related to SGs, and several applications were developed to show the advantages of VPPs. The authors of [48] proposed the integration of combined heat and power (CHP) microunits based on VPP in a low voltage network from a technical and economical point of view. The authors of [49] presented a new concept where microgrids and other production or consumption units form a VPP. The authors of [50] presented a concept VPP as a primary vehicle for delivering cost-efficient integration of distributed energy resources (DER) into the existing power systems. This study presented the technical and commercial functionality facilitated through the VPP and concluded with case studies demonstrating the benefit of aggregation and the use of the optimal power flow algorithm to characterize VPPs. The authors of [51] proposed the concept of generic VPP (GVPP), showing three case scenarios and overcoming challenges using a proposed solution framework and service-oriented architecture (SOA) as a technology which could aid in the implementation of GVPP. The authors of [52] provided a suitable software framework to implement GVPP with SOA. The FENIX European Project [53] delved into the concept of VPP and considered two types of VPP: the commercial VPP (CVPP) that tackles the aggregation of small generating units with respect to market integration and the technical VPP (TVPP) that tackles aggregation of these units with respect to services that can be offered to the grid. The authors of [54] described a general framework for future VPP to control low and medium voltage for DER management. The authors of [55] presented a case study which shows how a broker GVPP was developed based on the selection of appropriate functions. The

EDISON Danish project [56] described an ICT-based distributed software integration based on VPPs and standards to accommodate communication and optimize the coordination of EV fleets. The authors of [57] proposed an architecture for EV fleet coordination based on V2G integrating VPP. The authors of [58] analyzed the possibility of using EVs as an energy storage system (V2G) within a VPP structure. The authors of [59] considered the EV as a mobile load and described a VPP containing aggregated microgeneration sources and EV, but it is centered around minimizing carbon emissions. The authors of [60] proposed and discussed three approaches for grid integration of EVs through a VPP: control structure, resource type, and aggregation. The authors of [61] presented a solution for integrating EVs in the SG through unbundled smart metering and VPP technology dealing with multiple objectives. The authors of [62] addressed the design of an EV test bed which served as a multifunctional grid-interactive EV to test VPP or a generic EV coordinator with different control strategies.

The common point of these references is the utilization of the VPP concept in a simulation, but they only simulate the VPP which aggregates the information of EV. Although the aggregation idea is not always implemented with a VPP, for example, the authors in [63] proposed an aggregate battery modeling approach for EV fleet, which is aimed for energy planning studies of EV-grid integration, they did not use the concept of VPP. The present paper proposes the charging prioritization of EV fleets to provide additional services [64], like EV fleet management or demand forecasting.

Of course, there are not a lot of examples of EV fleet currently with a complete charging infrastructure, notwithstanding several references papers in the simulation focused in EV fleet simulation, for example, an evolutionary approach is proposed in Ref. [65] or a planning simulation model is presented in Ref. [66] that evaluates the feasibility of electric vehicle driving range when recharging is considered at home, at work, or at quick charging stations. But some scenarios are more difficult to simulate, for example, the use of electric modules which can be added or removed from a freight vehicle proposed in Ref. [67]. However, the problem, in this case, is the recharging of electric modules, which is done in different nodes or point of reception, and they will have a charging infrastructure; the authors provide a good mathematical background to calculate the time windows of electric module availability. This problem is similar to battery exchange infrastructures proposed in Ref. [68].

Additionally, a charging management could provide a good contribution to the demand forecasting, although the different references did not treat the problem of demand forecasting, but the scheduling and suitable assignment of EVs to charging stations could provide information about the demand in the charging stations. For example, the optimal solution provided in Ref. [69] could be a contribution for an algorithm to provide an aggregated demand forecasting. Other solutions are oriented to specific sectors or infrastructures: EV fleet parking determining the minimum number of chargers that are required to charge all electric vehicles [70] or estimating total daily impact of vehicles aggregated in parking lots on the grid [71], taxi fleets [72], and taxi fleets with mixed electric and conventional vehicles [73].

On the other hand, some researchers have studied the impact of HEV and plug-in HEV (PHEV) [74]. In this sense, decentralized algorithms for coordinating the charging of multiple EVs have gained importance in recent years. The authors of [75] compared several approaches based on centralized, decentralized, and hybrid algorithm, with the latter showing better results. The authors of [76] introduced the electric fleet size and mix vehicle routing problem with time windows and recharging stations (E-FSMFTW) to model decisions to be made with regard to fleet composition and vehicle routes, including the choice of recharging times and locations. The authors of [64] presented a review and classification of methods for smart charging of EVs for fleet operators, providing three control strategies and their commonly used algorithms. Additionally, they studied service relationships between fleet operators and four other actors in SGs.

3. Virtual power plants

The viewpoint of the proposed solution treats vehicles as a mobile load. In this manner, the system must have data about these loads and the charging prioritization. Thus, the system will have information about the expected consumption or the expected generation of the resource (in the case of a fault in the grid), such as a battery.

The proposed system works like a service for large companies with EV fleets. The knowledge about the state and prioritization of vehicles may minimize the impact of charging loads. These services provide new tariffs for retailers and new policies for energy price management.

The conceptual architecture of the proposed solution is shown in **Figure 1**. The proposed architecture is based on VPP concept [77]. Several VPPs are included. The information is aggregated on the lower level. Then, the aggregated information is sent by each lower VPP to a higher level. In this way, each VPP aggregates the data and services from lower VPPs to higher VPPs. Each level may have one or more VPPs, depending on the needs at each level and the power grid.

The information representation in different levels was based on an extension of the Common Information Model (CIM) from IEC 61970, 61,968, and 62,325. The interface information is based on the Component Interface Specification (CIS) from the IEC. The Open Automated Demand Response (OpenADR) version 2.0 is included in the VPP, but it is only enabled in some levels. The information representation and interface description are beyond the scope of this paper.

Each higher VPP can perform evolutionary algorithms to generate commands or instructions to modify the queues from lower VPPs. Additionally, lower VPPs can perform the same evolutionary algorithms to request resources from other VPPs to prioritize the charging of vehicles that cannot be charged at their charging stations.

4. The distributed evolutionary prioritization framework

The distributed evolutionary prioritization framework (DEPF) is implemented in each VPP. The architecture of this framework is shown in **Figure 2**. The modules are shown in **Figure 2**. Each module has specific functions:

- *Asset management system*. The asset management system is based on the predictive maintenance of vehicles and charging stations. These modules establish the maintenance periods and register the usage of all equipment (vehicles and charging stations).

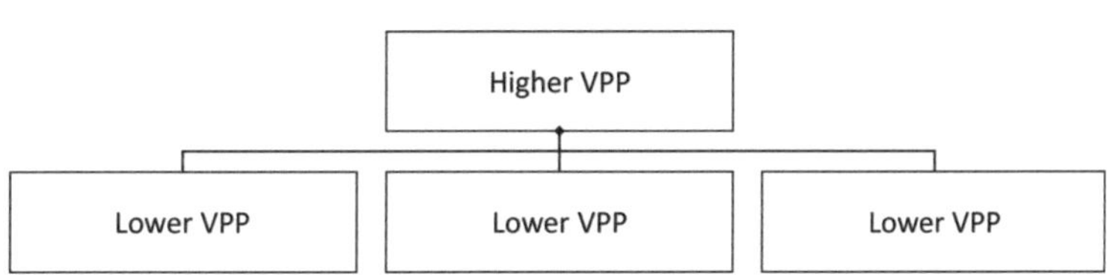

Figure 1.
Conceptual model of VPP.

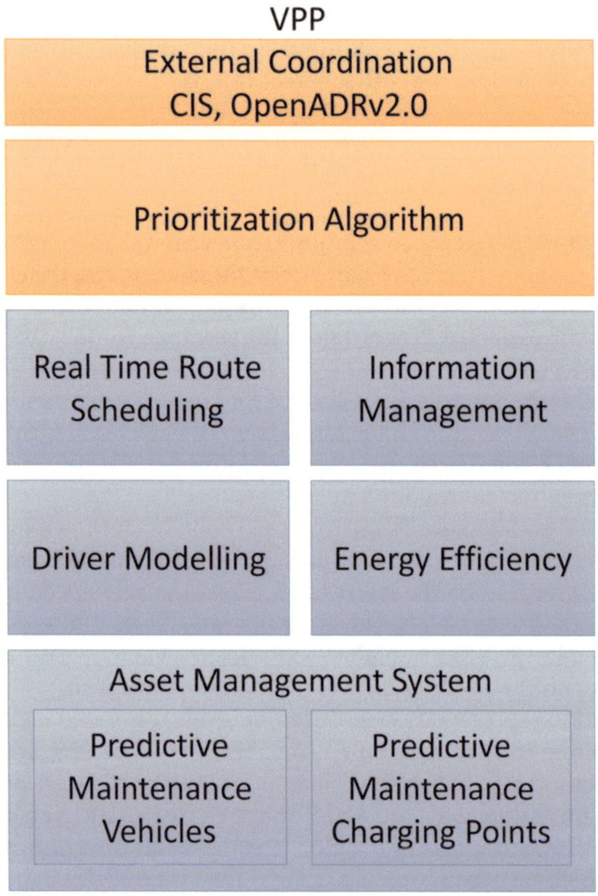

Figure 2.
Modules of distributed evolutionary prioritization framework.

- *Driver modeling*. This module executes a modeling process of driver behavior. This module provides a driver pattern, which is used to schedule the routes.

- *Energy efficiency*. This module applies different techniques to optimize the energy consumption and reduce the maintenance periods and economic impact.

- *Real-time route scheduling*. This module manages all information about vehicles, routes, drivers, and external conditions to establish better prioritization in each charging station.

- *Information management*. This module manages all information of this VPP for reporting and visualization.

- *Prioritization algorithm.* The prioritization algorithm in this layer is based on swarm intelligence.

- *External coordination*. This module sends information to higher layers and gathers information about external requirements or vehicles to charge. This module oversees communications with other VPPs (higher or lower) by using CIS or OpenADR.

Some modules, such as external coordination, prioritization algorithm, and the SoC module, are available for all VPPs. The other modules are enabled depending on the available information in the VPPs.

The modules shown in blue are not included in this chapter. The asset management system is in development and will not be described in this chapter.

The prioritization algorithm includes different options of evolutionary computations, to test the best algorithm. Several approaches are tested: genetic algorithm, genetic algorithm with evolution control, particle swarm optimization, and a hybrid solution. All these algorithms use the information available from the different modules. The information in each module is described in Sections 4.1 to 4.4. The information is channeled through information management module.

4.1 Available information

Much information about different entities is available, depending on the level of VPP. This information is used to calculate the SoC, the best driver for each router, and the prioritization at the charging stations of the company. When the route needs an external charging station (out of company supply), the final prioritization is assigned by higher VPP levels.

The available information is stored in a relational database management system (RDBMS) that is based on the IEC CIM. This information includes the following data: information about asset management, current configuration for prioritization algorithm, technical information about charging stations, parametric information about technical characteristics of different connection types for vehicles and charging stations, information about driver patterns, different measures gathered from charging stations and power register, establishing the expected periods of availability and nonavailability for each vehicle and charging station, information about pending and assigned routes, technical information about vehicles and their batteries, and information about traffic, roadwork, and topology. In addition, stored information about previously calculated charging prioritization and real information about charging stations. This information is stored to determine the difference between the expected charging process and the real charging process. This information will be useful to improve models for charging prioritization. Furthermore, other historical information included historical information about usage of charging stations, configuration of prioritization, drivers that will be used in the driver pattern modeling, traffic related to roadworks, weather conditions, traffic conditions and accidents, periods configured in the system, different routes stored in the system, statistical information, mechanical problems and statistics from EVs, and the execution, configuration, and results of configurations.

4.2 Driver patterns

Driver behavior is stored in driver patterns. The driver pattern is a model that takes effect on the consumption of a vehicle in route scheduling. The driver pattern affects the calculated SoC calculated for each section of a route; it depends on the terrain topology and traffic data. Driver behavior is calculated according to the historical data of a driver. If historical information about a driver is not available, this pattern cannot be calculated.

The driver pattern consists of the deviation from the original predicted SoC. This pattern considers information about traffic and weather to explain the variation from the original predicted SoC.

Although a default driver pattern can be defined, information about driver patterns is currently unavailable. A default "average" driver pattern can be created when a system has adequate information.

4.3 Real-time routing scheduling

This module controls several conditions that can modify the current prioritization charging queues. This module notifies any change in the following conditions: charging station availability, driver and EV availability, route modifications, traffic and roadworks, and weather conditions.

The charging station availability updates the unavailability periods of the corresponding charging stations, in this case, if the higher VPPs can send commands or instructions to limit the consumption or availability of the power supply.

On a route, the driver and EV availability is related to a driver (or EV). The driver may have an accident or a driver (or EV) may notify temporary unavailability. In this case, the module updates the calendar for the driver (or EV). This new condition takes effect in the prioritization process by a fitness function of evolutionary algorithms. Sometimes, the temporal unavailability of an EV can be notified by the asset management module.

Three types of route modifications are possible: adding new stop, adding new stop to charge an EV, and adding new stop to change drivers.

The traffic and work road modeling is translated into a penalty coefficient for different sections of a route. In this case, the route scheduling algorithm may provide the route with lesser penalties. These penalties take effect on the calculated SoC and the driver selection. The penalty coefficient is stored for each geographical zone and is associated with driver information. If a driver has a very high deviation from the original predicted value, the penalty is increased. This pattern did not consider any information about the origin of the traffic load. The pattern only assigns a penalty coefficient according to the fluidity of traffic.

The weather conditions would take effect over the SoC module, prioritization algorithm, and asset management module.

4.4 SoC module: estimation of EV consumption

The proposed solution is based on the SoC instantaneous value of each EV. These algorithms require an estimation of some consumptions, according to its planned route and alternative routes to achieve different recharging spots. This consumption estimation is supported by a route planning tool. However, these estimations are not trivial and relate to the distance or time of the trip [91, 92]. Other factors (e.g., road [78] and vehicle characteristics, traffic [79], driving style [80], and weather conditions) are essential for this estimation.

Typical approaches to estimating route consumptions must be reviewed and briefly explain the architecture that supports the main algorithm in this chapter. The two approaches can be easily distinguished in the literature:

- Knowledge-based models are the most common approach. This type of model performs a consumption characterization based on records of vehicle operations using computational intelligence techniques, such as artificial neural networks (ANNs) [81] and fuzzy neural networks (FNNs) [82]. However, these techniques have the disadvantage of requiring a large amount of data, which must contain different conditions to model realistic vehicle behaviors in as many situations as possible.

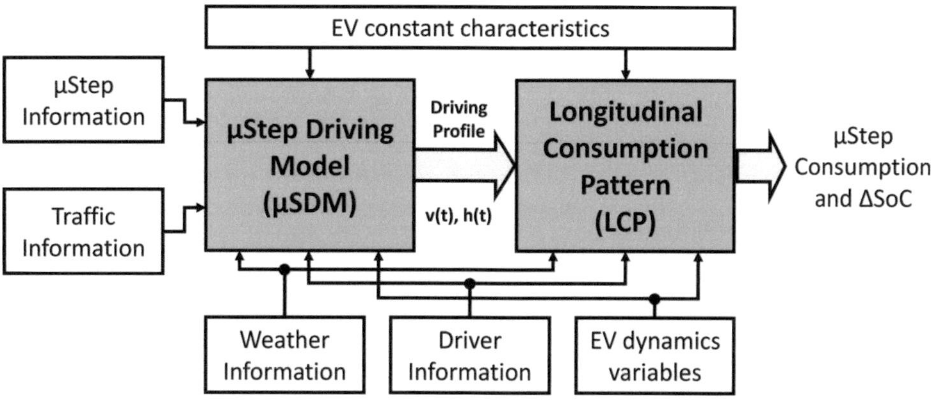

Figure 3.
Consumption model architecture.

- They are analytical models (also known as longitudinal models) that study the necessary energy to move a vehicle by analyzing the losses along its different mechanical and electronic elements [43, 93]. Thus, these models are typically more complex than knowledge-based models. However, analytical models do not require as much information as knowledge-based models because their parameters can be characterized based on information provided by a manufacturer (i.e., New European Driving Cycle (NEDC) standards test) [83].

Based on these philosophies and due to the lack of availability of necessary sets of registered data (especially in the initial phases of the project), an analytical model was chosen to estimate the consumption of this system. It provides the different SoCs that are associated with each route stage. This estimation is based on a model that can be divided into two blocks (refer to **Figure 3**): the first block—the μStep Driving Model (μSDM)—is responsible for estimating the driving profile. This estimate consists of the velocity and elevation profiles accomplished by the analyzed EV. The velocity profile is constructed from the averaged inferred patterns for different situations, which are characterized by the different input parameters. The elevation profile is directly estimated from the route information (in each μStep). A complete list of parameters in the model is shown in **Table 1**.

Parameter	Group	Associated to
μStep distance	μStep information	μSDM
μStep start elevation		μSDM
μStep final elevation		μSDM
μStep estimated time		μSDM
Traffic congestion level	Traffic information	μSDM
Driver style	Driver information	μSDM, LCP
Precipitation level	Weather information	μSDM, LCP
Average temperature		μSDM, LCP
Sunrise/sunset time		μSDM, LCP
ECO mode state	EV dynamics variables	μSDM, LCP
Additional mass (passengers, baggage, etc.)		LCP
Initial SoC		LCP

Parameter	Group	Associated to
Vehicle mass	EV constant characteristics	LCP
Eq. mass of rotating parts		LCP
Maximum acceleration		µSDM, LCP
Frontal area		LCP
Friction resistance coefficient		LCP
Drag coefficient		LCP
Performance internal systems (inverter + engine + transmission)		LCP
A/C consumption		LCP
Heating consumption		LCP
Cooling consumption		LCP
Light consumption		LCP
Power steering consumption		LCP
Secondary consumptions (light panels, control systems, etc.)		LCP
Battery performance		LCP
Battery capacity		LCP
Battery voltage		LCP
Maximum recovery capacity		LCP

Table 1.
List of consumption model inputs.

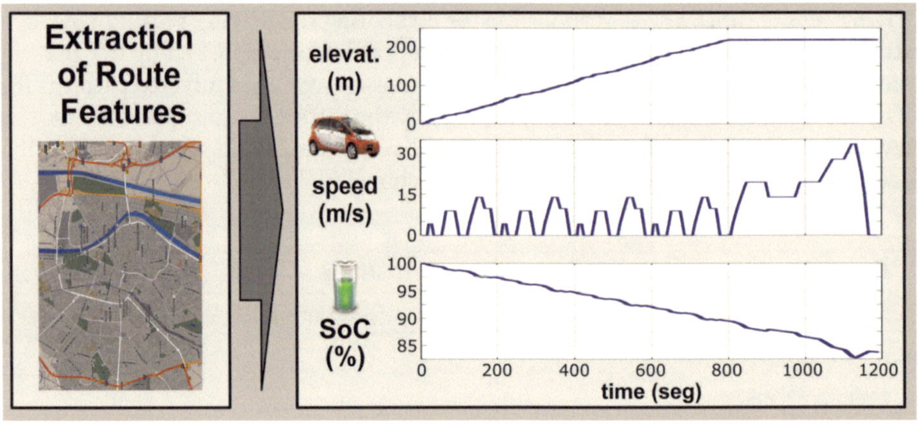

Figure 4.
Consumption and SoC estimation for a route.

The second block of the model—the longitudinal consumption pattern (LCP)—is responsible for estimating the consumptions and SoC variations associated with the velocity and elevation profiles for each µStep. This consumption is obtained by adding the different losses and consumptions that are required to generate these profiles, which are associated with the trip. This analysis also considers the vehicle configuration, as listed in **Table 1**.

The combination of part of the model and the initial SoC value enables the estimation of the SoC at the end of a trip or part there of (as shown in **Figure 4**).

5. Prioritization algorithm

Three algorithms are proposed to prioritize the charging queues:

- Genetic algorithm (GA)
- Genetic algorithm with evolution control (GAEC) based on fitness evolution curve
- Swarm intelligence based on particle swarm optimization (PSO)

These algorithms were applied in the same scenario and different layers to determine the best combination of algorithms.

The application of these algorithms is performed after a preprocessing stage after the following steps:

1. The routes are sorted by starting timestamp, route distance, and ending timestamp.
2. The vehicles are sorted by range or battery capacity.
3. The drivers are sorted according to the difference between the real and expected routes, in ascending order. The new drivers or drivers without historical information are positioned in first place.
4. The charging stations are ordered by connector type, according to the estimated charging period.

5.1 Genetic algorithm (GA)

The genetic algorithm (GA) [94, 95] is a bioinspired algorithm. This algorithm is based on the evolution of populations, in which only the best individuals survive. Everyone from a population is a possible solution to a problem, and a fitness value is assigned according to an indicator that determines the distance from the final solution. In each evolution, a new generation from the previous population is created based on cross, mutation, and selection processes. After some evolutions (iterations), the algorithm converges to a best solution or a solution that complies with the threshold.

Algorithm 1. Genetic algorithm (GA): size p of population (P(t)), rate q of elitism, rate c of crossover (default 0,9), and rate m of mutation (default 0,1).

---

1. Randomly generate p feasible solutions,
2. Save them in the population P(t),
3. Evaluation of the population P(t); thus, the fitness of each solution of the population P(t) are determined.
4. Repeat
    4.1. Select parents from the population P(t), number of elitism n_e=p*q
    4.2. Perform the crossover on parents by creating the new population P(t+1) with the probability c.
    4.3. Perform mutation of the population P(t+1) with the probability m.
    4.4. Assess the population P(t+1).
    4.5. If the stopping criteria are true, then return to step 3; otherwise, proceed to step 5.
5. If the threshold is active, then obtain all solutions for which the fitness value complies with the threshold; otherwise, obtain the best solution (the best evaluation value).

The stopping criteria are specified in the configuration of the prioritization algorithm.

5.2 Genetic algorithm with evolution control (GAEC) based on fitness evolution curve

The GAEC is a genetic algorithm with additional restrictions that influence the probability of mutate and cross operators according to the fitness evolution curve. The fitness evolution curve is created in each evolution and stores the best fitness value of each evolution. The angle of the tangent line in each point of the fitness evolution curve determines the probability of the operators.

Algorithm 2. Genetic algorithm with evolution control (GAEC): size p of the population (P(t)), rate q of elitism.

---

1. Randomly generate p feasible solutions,
2. Save them in the population P(t),
3. Evaluation of the population P(t); thus, the fitness of each solution of the population P(t) are determined.
4. Repeat
   - 4.1. Select parents from the population P(t), number of elitism $n_e = p^*q$
   - 4.2. Perform the crossover on parents by creating the new population P(t + 1) with the probability based on the absolute value of the sine of the angle of the tangent line to the fitness evolution curve.
   - 4.3. Perform a mutation of the population P(t + 1) with the probability based on the absolute value of the cosine of the angle of the tangent line to the fitness evolution curve.
   - 4.4. Assess the population P(t + 1).
   - 4.5. If the stopping criteria are true, then return to 3; otherwise, proceed to 5.
5. If the threshold is active, then obtain all solutions for which the fitness value complies with the threshold; otherwise, obtain the best solution (the best evaluation value).

5.3 Particle swarm optimization (PSO) algorithm

The prioritization algorithm works as a swarm intelligence algorithm. The application of the algorithm is performed after a preprocessing of information:

The prioritization algorithm is based on the parametric optimization until a solution is obtained. This optimization is executed depending on the capabilities of a system. The algorithm employs a PSO to establish the initial prioritization for the charging stations in the company area. The canonical PSO model consists of a swarm of particles, which are initialized with a population of random candidate solutions. They iteratively move through the d-dimension problem space to search for the new solutions, where the fitness f can be calculated as the certain quality measure. Each particle has a position that is represented by the position-vector x_{id} (i is the index of the particle, and d is the dimension) and a velocity represented by the velocity-vector vid. Each particle remembers its best position in the vector $x_{i\#}$, and its j-th dimensional value is $x_{\#ij}$. The best position-vector among the swarm is stored in the vector x_*, and its j-th dimensional value is x_{*j}. At the iteration time t, the update of the velocity from the previous velocity to the new velocity is determined by Eq. (1). The new position is determined by the sum of the previous position, and the new velocity is determined by Eq. 2.

$$v_{id}(t+1) = w * v_{id}(t) + c_1 * \psi_1 * (p_{id}(t) - x_i(t)) + c_2 * \psi_2 * \left(p_g(t) - x_{id}(t)\right) \quad (1)$$

$$x_{id}(t+1) = x_{id}(t) + v_{id}(t+1) \quad (2)$$

where c_1 and c_2 are constant weight factors, pi is the best position achieved by particle i, p_g is the best position obtained by the neighbors of particle i, ψ_1 and ψ_2 are random factors in the [0,1] interval, and w is the inertia weight. Some references denote c_1 and c_2 as the self-recognition component and the coefficient of the social component, respectively.

Different constraints can be applied to ensure convergence of the algorithm.

Algorithm 3. PSO Algorithm

1. Initialize particles
2. Repeat
    - 2.1. Calculate the fitness values for each particle
    - 2.2. Is the current fitness value better than p_i?
        - 2.2.1. Yes. Assign the current fitness as the new p_i
        - 2.2.2. No. Keep the previous p_i
    - 2.3. Assign the best particle's p_i value to p_g
    - 2.4. Calculate the velocity for each particle
    - 2.5. Use each particle's velocity value to update its data values
3. Until stopping criteria are satisfied.

5.4 Configuration of prioritization algorithms

The system enables different restrictions to be specified for the prioritization algorithm: assignation prioritization of external charge, driver rest periods along a route, driver rest periods between different routes, external charging priority, external charging, maintenance periods for charging stations, maintenance periods for vehicles, possibility of partial charging, possibility of reuse drivers, possibility of several vehicles per route, possibility of specifying periods of unavailability of charging stations, possibility of specifying periods of unavailability of drivers, possibility of specifying periods of unavailability of vehicles, rest periods between vehicles that charge at charging stations, time interval to prioritize (1 day by default), and usage balancing of charging stations.

The external charging priority takes effect in the way which system will assign the first available slot in the queues to the external vehicles; however, the system moves the vehicles of the lowest VPPs (if possible). Furthermore, the system may accept a charging request from the different VPPs.

These parameters can be modified while the algorithm is running. These parameters take effect over the convergence of evolutionary algorithms because the parameters can modify the fitness function.

The evolutionary algorithms are based on an iterative algorithm. In this case, the proposed algorithms have several similarities. These algorithms have an end criteria to control the iterative part of the algorithm. In the proposed algorithms, the end criteria can be configured by the user by specifying one or more parameters:

- Maximum number of iterations: the optimization process is terminated after a fixed number of iterations.

- Number of iterations without improvements: the optimization process is terminated after a fixed number of interactions without any iterations and without any improvement.

- Minimum objective function error: the error between the obtained objective function value and the best fitness value is less than a prefixed anticipated threshold.

The value of these parameters is dependent on the size of the search space and the complexity of the problem. These values are established by the default value according to the number of vehicles, number of drivers, number of routes, and number of charging stations, as well as the system characteristics in the installation stage. Other parameters in the PSO algorithm are dependent on the same parameters: maximum number of particles (swarm size or number of neighbors), maximum velocity (v_{max}) for the PSO algorithm [84], maximum particle position ($x_{i,max}$) for the PSO algorithm [84] to retain the value of the particle position in the interval $[-x_{i,max}, x_{i,max}]$, inertia weight (w) for PSO algorithm, modifiers for random number generation, self-recognition component (must be a positive value), coefficient of the social component (must be a positive value), and maximum and minimum velocities.

A GA and GAEC have similar parameters: population size, and, only in the GA, the operator probabilities (mutation and crossover).

These factors are employed in the algorithm to fix the evolution of each particle and dimension, in the case of PSO, or everyone in a population, in the case of a GA and GAEC. The previously defined parameters and the parameters defined in this section can modify the convergence of an algorithm. These parameters are automatically adjusted in each evolution and running.

5.5 Fitness function

The fitness function is calculated to test the validity of a particle. The fitness function is based on the following items: the number of routes that have been assigned to a vehicle and driver for the time requirements to perform the route and the number of routes that are assigned to a vehicle and driver but exceeds the time requirements to perform the route are significantly penalized.

Additionally, the parameters that are configured in the system can modify the final fitness value that is calculated for each solution: queue balancing of a charging station, external charging, external charging priority, reuse of vehicles, reuse of drivers, assignation prioritization of external charging, several vehicles per route, instructions from higher VPPs, or presence of autonomous vehicles.

Several of these configurations can change at any time. Thus, the fitness value can change for each generation of evolutionary algorithm.

The proposed fitness function is performed in the proposed evolutionary algorithms. Thus, the fitness value is normalized in the interval [−1,1]. This fact simplifies the comparison of different options.

6. Experimental results

The proposed algorithm was tested in different scenarios. These scenarios were simulated using a computer. Several entities are created:

- Two smart business parks (A and B) with separate EV fleets. Company A is a company in the logistics sector, and company B is a company in the transport sector.

- Three public charging stations.

- Five private EVs.

The different levels of VPPs are defined as shown in **Figure 5**. Each level runs the DEPF. The VPP level at which the DEPF is performed determines the availability of services, protocols, and data. Several VPP levels are proposed (**Figure 5**):

- *Smart business VPP (SBVPP).* This is the lowest level. At this level, all information about vehicles, routes, and drivers from the same company is available. Thus, the charging prioritization of the charging stations of the company is treated at this level. The state of charge (SoC) is also calculated at this level. Some routes may be very long, which may cause a vehicle to use a charging station that is located outside of the company. This charging station may be administered by another company or the corresponding power distribution company. In this case, the algorithm sends the restrictions to higher VPP levels to obtain a solution for the charging needs. This VPP communication is based on CIS and OpenADR protocols.

- *Distribution VPP (DVPP).* At this level, information is aggregated from lower levels, and information about retailers and the presence of charging stations is stored. This information is sent to higher levels, such as an energy VPP (EVPP). Additionally, the restrictions from an EVPP to the corresponding retailer and SBVPP are addressed at this level. This VPP communication is based on CIS and OpenADR protocols.

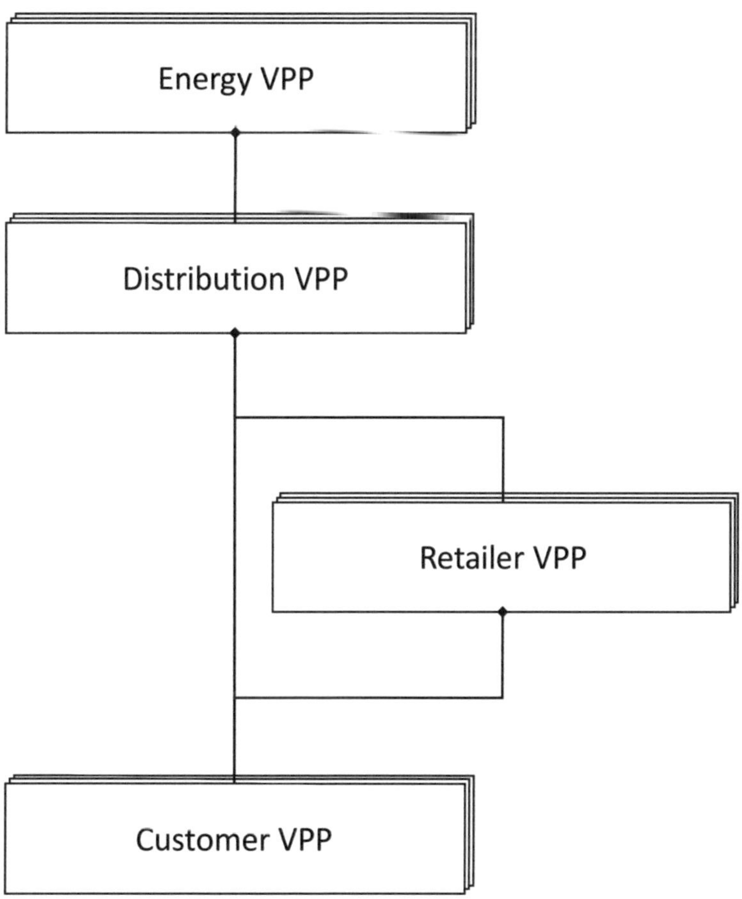

Figure 5.
Information aggregation between different VPP layers proposed.

- *Retailer VPP (RVPP).* At this level, a retailer needs to know when vehicles require charging at any point outside of the company points. The retailer can use this information to offer different tariffs to a company. This level acts as an intermediate between charging stations of different companies. This VPP communication is based on OpenADR protocol.

- *Energy VPP (EVPP).* In this paper, the vehicles represent mobile loads. Thus, if an energy management system has information about the expected charging stations, it may take advantage of this information to improve the load flow forecasting algorithms. The load flow forecasting algorithm is not an objective of this paper. This paper proposes a distributed prioritization algorithm based on the VPP concept for SGs. The prioritization algorithm that is performed at this level treats the total load and establishes possible restrictions at any point of the grid. This VPP communication is based on CIS and OpenADR protocols.

From the point of view of the power market:

- Two power retail companies. The first retailer has a contract to supply company A. The second retailer has a contract to supply company B. The retailer has three contracts to supply private consumers.

- One power distribution company.

- An EMS is simulated. This system is configured to randomly generate a power consumption command in the EVPP. This power consumption command takes effect on 171 routes: 68 routes from company A and 103 routes from company B. This power consumption will be generated after a solution is obtained to assess the algorithm and address any changes in conditions.

- Companies A and B have an SBVPP. The characteristics of these companies are listed in **Table 2**. For each driver, EV, and charging station, some periods of unavailability are defined to check the capability of the algorithms to manage these contingencies.

- In this case, the private consumers are managed by the RVPP.

The evaluation of the proposed solution is conducted in several scenarios based on algorithms: GA, GAEC, PSO, and hybrid solution. In case of hybrid solution, all possible combinations (81 cases) were tested; however, only the best hybrid

Characteristics	Company A	Company B
Number of routes	200	300
Number of EVs	4	7
Number of drivers	3	6
Number of charging stations	2	2
Number of plugs by charging station	2	3
Power of charging stations	DC 50 kW/AC 43 kVA	DC 50 kW/AC 43 kVA
Time of fast charging (0–80%)	30 minutes	30 minutes

Table 2.
Characteristics of both companies with EV fleets.

solution is shown in this chapter. The best hybrid solution applies PSO in SBVPP and GAEC in the higher VPPs.

The proposed solution is evaluated by checking two aspects (**Table 3**); both aspects are evaluated in the general best fitness curve:

- *Convergence time (t_c).* Time to reach a solution. The convergence time is measured in number of generations.

- *Transient time (t_t).* Time to obtain a new solution when changes occur in the conditions of the problem. The transient time is measured in a number of generations.

The general best fitness curve for each scenario is shown in **Figure 6**. The first part (until generation 17) of the curve is the search of the best solution to schedule the routes (500 routes). The corresponding number of scheduled routes is shown in **Figure** 7. After the best solution is obtained, the simulated EMS randomly generates several commands in each scenario. In the case of the GA scenario, the command was fired in generation 34 (**Figure 6**) with 258 scheduled routes (**Figure** 7); in the GAEC scenario, the command was fired in generation 19 with 456 scheduled routes; in the PSO scenario, the command was fired in generation 30 with 338 scheduled routes; and in the hybrid scenario, the command was fired in generation 39 with 378 scheduled routes. The command takes effect in a different number of routes in each scenario; the results did not agree with the fitness value (**Figure 6**) and the number of scheduled routes (**Figure** 7) between different scenarios, because there are different scheduling solutions. When the command is fired, the fitness is updated with

Test scenarios	t_c (number of generations)	t_t (number of generations)
Only GA	16	5
Only GAEC	11	2
Only PSO	17	5
Hybrid solution	11	3

Table 3.
Evaluation parameters for each scenario.

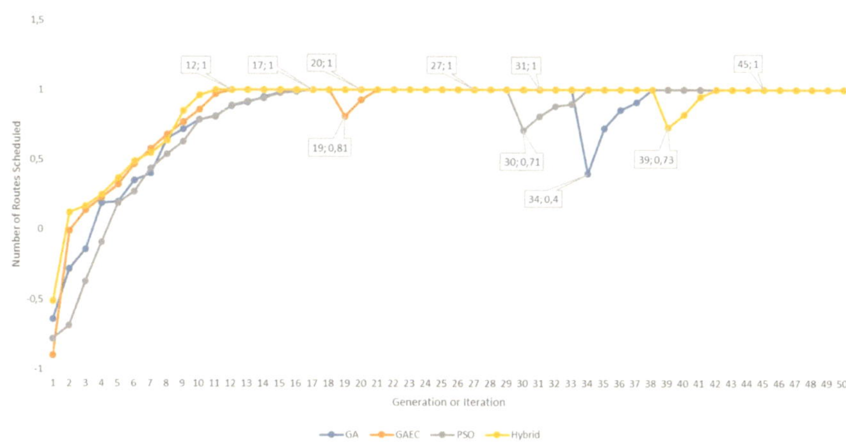

Figure 6.
General best fitness curve.

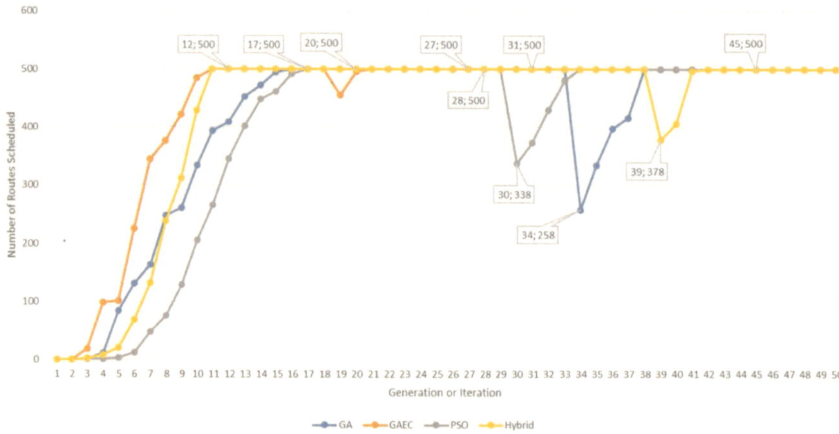

Figure 7.
Translate of general best fitness curve to the number of assigned routes.

the new restrictions. This updating changed the value of the best solution by converting it into another solution with lower fitness. However other solutions in the same population generation could have a good fitness in the new scenario. In the population (in the case of GA and GAEC) or swarm (in the case of PSO), several solutions were reassessed. Although some solutions were not the best solutions in the initial scenario, after the command was fired, the new fitness value improved, and the solutions were sorted according to this new assessed fitness.

In the same manner, the transient time is less than the initial part of the algorithm because the current solutions are disseminated in the space of possible solutions. Thus, the solution is obtained at a faster rate.

According to the results, some conclusions are formed:

- GA and PSO exhibit the best trend.
- GAEC performs better in transient situations.
- The hybrid solution obtains better results because it takes advantage of all evolutionary algorithms.

One of the most interesting effects is shown in **Figure 6** in GAEC (fitness GAEC) scenario. This algorithm has several steps. In these steps, the mutation probability is increased, and the crossover is decreased. This fact disseminates the solutions in the space of possible solutions, which increases the probability of obtaining better solutions.

7. Conclusions

A novel solution for the distributed prioritization of charging station queues is presented in this chapter. The proposed solution provides additional results:

- An algorithm to manage the EV fleet, to improve the efficiency of fleet.
- A model of mobile load inside a power grid. The algorithm provides a load forecasting of mobile loads, and it again calculates in real time in case an unexpected incident or an additional EV is added.

- The comparison between different energy management scenarios showing the hybrid solution as the best solution to different scenarios.

- The aggregation in different levels decreases the response time of system at different levels, allowing to respond in real time. In the consumer level, the EV is charging with minimized waiting periods. In the retailer level, the retailer can offer different rates and services according to the demand forecasting. In the distribution and energy level, the asset management, the energy flow, and the demand peak shaving are simplified based on demand forecasting.

- The reduction of waiting time to charge the EV. The prioritization takes into account the minimization of the waiting time.

- The successful usage of CIM and CIS in a VPP-based environment.

Although the usage of EVs can be an excellent solution for decreasing pollution, it may cause serious problems in the power grid. Several solutions could be applied to solve this problem. In this chapter, the proposed solution is to establish prioritization queues that enable control of the mobile loads or EV charging by taking advantage of the fact that this EV can only be plugged into charging stations. This type of knowledge can help energy management systems and other participants of power distribution to maintain a high quality of service and supply. Additionally, this knowledge provides information for distributed energy resource systems in the case of an alarm or emergency; in this case, the battery of EV (V2G) can serve as an energy resource.

Author details

Juan Ignacio Guerrero Alonso*, Enrique Personal, Antonio Parejo, Sebastián García, Antonio García and Carlos León
Department of Electronic Technology, Escuela Politécnica Superior (EPS), Universidad de Sevilla, Seville, Spain

*Address all correspondence to: juaguealo@us.es

References

[1] Boulanger AG, Chu AC, Maxx S, Waltz DL. Vehicle electrification: Status and issues. Proceedings of the IEEE. 2011;**99**(6):1116-1138

[2] Mahmud K, Town GE, Morsalin S, Hossain MJ. Integration of electric vehicles and management in the internet of energy. Renewable and Sustainable Energy Reviews. 2018;**82**: 4179-4203

[3] Fernandez LP, Roman TGS, Cossent R, Domingo CM, Frias P. Assessment of the impact of plug-in electric vehicles on distribution networks. IEEE Transactions on Power Systems. 2011;**26**(1):206-213

[4] Lopes JAP, Soares FJ, Almeida PMR. Integration of electric vehicles in the electric power system. Proceedings of the IEEE. 2011;**99**(1): 168-183

[5] Sadeghi-Barzani P, Rajabi-Ghahnavieh A, Kazemi-Karegar H. Optimal fast charging station placing and sizing. Applied Energy. 2014;**125**: 289-299

[6] Fiori C, Ahn K, Rakha HA. Power-based electric vehicle energy consumption model: Model development and validation. Applied Energy. 2016;**168**:257-268

[7] Rezvanizaniani SM, Liu Z, Chen Y, Lee J. Review and recent advances in battery health monitoring and prognostics technologies for electric vehicle (EV) safety and mobility. Journal of Power Sources. 2014;**256**: 110-124

[8] Sortomme E, Hindi MM, MacPherson SDJ, Venkata SS. Coordinated charging of plug-In hybrid electric vehicles to minimize distribution system losses. IEEE Transactions on Smart Grid. 2011;**2**(1): 198-205

[9] Karfopoulos EL, Hatziargyriou ND. Distributed coordination of electric vehicles providing V2G services. IEEE Transactions on Power Systems. 2016; **31**(1):329-338

[10] Shayegan-Rad A, Badri A, Zangeneh A. Day-ahead scheduling of virtual power plant in joint energy and regulation reserve markets under uncertainties. Energy. 2017;**121**: 114-125

[11] Qian K, Zhou C, Allan M, Yuan Y. Modeling of load demand due to EV battery charging in distribution systems. IEEE Transactions on Power Systems. 2011;**26**(2):802-810

[12] López MA, de la Torre S, Martín S, Aguado JA. Demand-side management in smart grid operation considering electric vehicles load shifting and vehicle-to-grid support. International Journal of Electrical Power & Energy Systems. 2015;**64**:689-698

[13] Kempton W, Tomić J. Vehicle-to-grid power fundamentals: Calculating capacity and net revenue. Journal of Power Sources. 2005;**144**(1):268-279

[14] van der Kam M, van Sark W. Smart charging of electric vehicles with photovoltaic power and vehicle-to-grid technology in a microgrid; a case study. Applied Energy. 2015;**152**:20-30

[15] Jaguemont J, Boulon L, Dubé Y. A comprehensive review of lithium-ion batteries used in hybrid and electric vehicles at cold temperatures. Applied Energy. 2016;**164**:99-114

[16] Fathabadi H. Novel solar powered electric vehicle charging station with the capability of vehicle-to-grid. Solar Energy. 2017;**142**:136-143

[17] Uddin K, Jackson T, Widanage WD, Chouchelamane G, Jennings PA,

Marco J. On the possibility of extending the lifetime of lithium-ion batteries through optimal V2G facilitated by an integrated vehicle and smart-grid system. Energy. 2017;**133**:710-722

[18] Meissner E, Richter G. Battery monitoring and electrical energy management: Precondition for future vehicle electric power systems. Journal of Power Sources. 2003;**116**(1):79-98

[19] Tie SF, Tan CW. A review of energy sources and energy management system in electric vehicles. Renewable and Sustainable Energy Reviews. 2013;**20**: 82-102

[20] Li S, Mi CC. Wireless power transfer for electric vehicle applications. IEEE Journal of Emerging and Selected Topics in Power Electronics. 2015;**3**(1): 4-17

[21] Shareef H, Islam MM, Mohamed A. A review of the stage-of-the-art charging technologies, placement methodologies, and impacts of electric vehicles. Renewable and Sustainable Energy Reviews. 2016;**64**:403-420

[22] Divakarla KP, Emadi A, Razavi S. A cognitive advanced driver assistance systems architecture for autonomous-capable electrified vehicles. IEEE Transactions on Transportation Electrification. 2019;**5**(1):48-58

[23] Chau KT, Wong YS, Chan CC. An overview of energy sources for electric vehicles. Energy Conversion and Management. 1999;**40**(10):1021-1039

[24] Laberteaux KP, Hamza K. A study on opportune reduction in greenhouse gas emissions via adoption of electric drive vehicles in light duty vehicle fleets. Transportation Research Part D: Transport and Environment. 2018;**63**: 839-854

[25] Nelson RF. Power requirements for batteries in hybrid electric vehicles. Journal of Power Sources. 2000;**91**(1): 2-26

[26] Gerssen-Gondelach SJ, Faaij APC. Performance of batteries for electric vehicles on short and longer term. Journal of Power Sources. 2012;**212**: 111-129

[27] Martel F, Dubé Y, Kelouwani S, Jaguemont J, Agbossou K. Long-term assessment of economic plug-in hybrid electric vehicle battery lifetime degradation management through near optimal fuel cell load sharing. Journal of Power Sources. 2016;**318**:270-282

[28] Marra F, Sacchetti D, Træholt C, Larsen E. Electric vehicle requirements for operation in smart grids. In: 2011 2nd IEEE PES International Conference and Exhibition on Innovative Smart Grid Technologies (ISGT Europe). 2011. pp. 1-7

[29] Giannouli M, Yianoulis P. Study on the incorporation of photovoltaic systems as an auxiliary power source for hybrid and electric vehicles. Solar Energy. 2012;**86**(1):441-451

[30] Romero-Ruiz J, Pérez-Ruiz J, Martin S, Aguado JA, De la Torre S. Probabilistic congestion management using EVs in a smart grid with intermittent renewable generation. Electric Power Systems Research. 2016; **137**:155-162

[31] Ross SM. 8—Queueing theory. In: Ross SM, editor. Introduction to Probability Models. 12th ed. London, UK: Academic Press; 2019. pp. 507-589. Available from: http://www.sciencedirect.com/science/article/pii/B9780128143469000135

[32] Bachmat E. Airplane boarding meets express line queues. European Journal of Operational Research. 2019;**275**(3): 1165-1177

[33] Pourvaziri H, Pierreval H. Dynamic facility layout problem based on open

queuing network theory. European Journal of Operational Research. 2017; **259**(2):538-553

[34] Babicheva TS. The use of queuing theory at research and optimization of traffic on the signal-controlled road intersections. Procedia Computer Science. 2015;**55**:469-478

[35] Fedorenko V, Fedorenko I, Sukmanov A, Samoylenko V, Shlaev D, Atanov I. Modeling of data acquisition systems using the queueing theory. AEU —International Journal of Electronics and Communications. 2017;**74**:83-87

[36] Apte V. Performance analysis of distributed software systems: Approaches based on queueing theory. In: 2007 Working IEEE/IFIP Conference on Software Architecture (WICSA'07). 2007. p. 39

[37] Ilić D, Karnouskos S, Beigl M. Improving accuracy of energy forecasting through the presence of an electric vehicle fleet. Electric Power Systems Research. 2015;**120**:32-38

[38] Milligan R. 13—Drive cycles for battery electric vehicles and their fleet management. In: Muneer T, Kolhe ML, Doyle A, editors. Electric Vehicles: Prospects and Challenges. Amsterdam, Netherlands: Elsevier; 2017. pp. 489-555. Available from: http://www.sciencedirect.com/science/article/pii/B9780128030219000136 [Accessed: June 5, 2019]

[39] Burgess M, Harris M, Walsh C, Carroll S, Mansbridge S, King N. A longitudinal assessment of the viability of electric vehicles for daily use. In: IET Hybrid and Electric Vehicles Conference 2013 (HEVC 2013). 2013. pp. 1-5

[40] Yi Z, Smart J, Shirk M. Energy impact evaluation for eco-routing and charging of autonomous electric vehicle fleet: Ambient temperature consideration. Transportation Research Part C: Emerging Technologies. 2018;**89**: 344-363

[41] Farhan J, Chen TD. Impact of ridesharing on operational efficiency of shared autonomous electric vehicle fleet. Transportation Research Part C: Emerging Technologies. 2018;**93**: 310-321

[42] Škugor B, Deur J. A bi-level optimisation framework for electric vehicle fleet charging management. Applied Energy. 2016; **184**:1332-1342

[43] Carli R, Dotoli M. A distributed control algorithm for optimal charging of electric vehicle fleets with congestion management. IFAC-PapersOnLine. 2018;**51**(9):373-378

[44] Macrina G, Laporte G, Guerriero F, Di Puglia Pugliese L. An energy-efficient green-vehicle routing problem with mixed vehicle fleet, partial battery recharging and time windows. European Journal of Operational Research. 2019; **276**(3):971-982

[45] Seddig K, Jochem P, Fichtner W. Integrating renewable energy sources by electric vehicle fleets under uncertainty. Energy. 2017;**141**:2145-2153

[46] Li Z, Guo Q, Sun H, Xin S, Wang J. A new real-time smart-charging method considering expected electric vehicle fleet connections. IEEE Transactions on Power Systems. 2014;**29**(6):3114-3115

[47] Dielmann K, van der Velden A. Virtual power plants (VPP)—A new perspective for energy generation? In: Modern Techniques and Technologies, 2003 MTT 2003 Proceedings of the 9th International Scientific and Practical Conference of Students, Post-Graduates and Young Scientists. 2003. pp. 18-20

[48] Schulz C, Roder G, Kurrat M. Virtual power plants with combined heat and power micro-units. In: 2005

International Conference on Future Power Systems. 2005. p. 5

[49] Dimeas AL, Hatziargyriou ND. Agent based control of virtual power plants. In: International Conference on Intelligent Systems Applications to Power Systems, ISAP. 2007. pp. 1-6

[50] Pudjianto D, Ramsay C, Strbac G. Virtual power plant and system integration of distributed energy resources. IET Renewable Power Generation. 2007;**1**(1):10-16

[51] Andersen PB, Poulsen B, Decker M, Traeholt C, Ostergaard J. Evaluation of a generic virtual power plant framework using service oriented architecture. In: PECon 2008: 2008 IEEE 2nd International Power and Energy Conference. 2008. pp. 1212-1217

[52] Andersen PB, Poulsen B, Træholt C, Østergaard J. Using service oriented architecture in a generic virtual power plant. In: Sixth International Conference on Information Technology: New Generations, 2009 ITNG '09. 2009. pp. 1621-1622

[53] Kieny C, Berseneff B, Hadjsaid N, Besanger Y, Maire J. On the concept and the interest of virtual power plant: Some results from the European project Fenix. In: IEEE Power Energy Society General Meeting, 2009 PES '09. 2009. pp. 1-6

[54] Mashhour E, Moghaddas-Tafreshi SM. The opportunities for future virtual power plant in the power market, a view point. In: 2009 International Conference on Clean Electrical Power. 2009. pp. 448-452

[55] You S, Træholt C, Poulsen B. Generic virtual power plants: Management of distributed energy resources under liberalized electricity market. In: 8th International Conference on Advances in Power System Control, Operation and Management (APSCOM 2009). 2009. pp. 1-6

[56] Binding C, Gantenbein D, Jansen B, Sundström O, Andersen PB, Marra F, et al. Electric vehicle fleet integration in the danish EDISON project—A virtual power plant on the island of Bornholm. In: IEEE PES General Meeting. 2010. pp. 1-8

[57] Jansen B, Binding C, Sundstrom O, Gantenbein D. Architecture and communication of an electric vehicle virtual power plant. In: 2010 First IEEE International Conference on Smart Grid Communications (SmartGridComm). 2010. pp. 149-154

[58] Musio M, Lombardi P, Damiano A. Vehicles to grid (V2G) concept applied to a virtual power plant structure. In: 2010 XIX International Conference on Electrical Machines (ICEM). 2010. pp. 1-6

[59] Skarvelis-Kazakos S, Papadopoulos P, Grau I, Gerber A, Cipcigan LM, Jenkins N, et al. Carbon optimized virtual power plant with electric vehicles. In: 2010 45th International Universities Power Engineering Conference (UPEC). 2010. pp. 1-6

[60] Raab AF, Ferdowsi M, Karfopoulos E, Unda IG, Skarvelis-Kazakos S, Papadopoulos P, et al. Virtual power plant control concepts with electric vehicles. In: 2011 16th International Conference on Intelligent System Application to Power Systems (ISAP). 2011. pp. 1-6

[61] Sanduleac M, Eremia M, Toma L, Borza P. Integrating the electrical vehicles in the smart grid through unbundled smart metering and multi-objective virtual power plants. In: 2011 2nd IEEE PES International Conference and Exhibition on Innovative Smart Grid Technologies (ISGT Europe). 2011. pp. 1-8

[62] Marra F, Sacchetti D, Pedersen AB, Andersen PB, Traholt C, Larsen E.

Implementation of an electric vehicle test bed controlled by a virtual power plant for contributing to regulating power reserves. In: 2012 IEEE Power and Energy Society General Meeting. 2012. pp. 1-7

[63] Škugor B, Deur J, et al. Energy. 2015;**92**:444-455

[64] Hu J, Morais H, Sousa T, Lind M. Electric vehicle fleet management in smart grids: A review of services, optimization and control aspects. Renewable and Sustainable Energy Reviews. 2016;**56**:1207-1226

[65] Jäger B, Hahn C, Lienkamp M. An evolutionary algorithm for an agent-based fleet simulation focused on electric vehicles. In: 2016 International Conference on Collaboration Technologies and Systems (CTS). 2016. pp. 457-464

[66] Usman M, Knapen L, Yasar A-U-H, Bellemans T, Janssens D, Wets G. Optimal recharging framework and simulation for electric vehicle fleet. In: Future Generation Computer Systems. 2017. Available from: http://www.sciencedirect.com/science/article/pii/S0167739X17307689 [Accessed: June 5, 2019]

[67] Rezgui D, Chaouachi Siala J, Aggoune-Mtalaa W, Bouziri H. Application of a variable neighborhood search algorithm to a fleet size and mix vehicle routing problem with electric modular vehicles. Computers & Industrial Engineering. 2019;**130**: 537-550

[68] Mirchandani P, Adler J, Madsen OBG. New logistical issues in using electric vehicle fleets with battery exchange infrastructure. Procedia-Social and Behavioral Sciences. 2014;**108**:3-14

[69] Mkahl R, Nait-Sidi-Moh A, Gaber J, Wack M. An optimal solution for charging management of electric vehicles fleets. Electric Power Systems Research. 2017;**146**:177-188

[70] Álvaro R, Fraile-Ardanuy J. Charge scheduling strategies for managing an electric vehicle fleet parking. In: IEEE EUROCON 2015—International Conference on Computer as a Tool (EUROCON). 2015. pp. 1-6

[71] Rezaee S, Farjah E, Khorramdel B. Probabilistic analysis of plug-In electric vehicles impact on electrical grid through homes and parking lots. IEEE Transactions on Sustainable Energy. 2013;**4**(4):1024-1033

[72] Dutta P. Charge sharing model using inductive power transfer to increase feasibility of electric vehicle taxi fleets. In: 2014 IEEE PES General Meeting|Conference Exposition. 2014. pp. 1-4

[73] Lu C-C, Yan S, Huang Y-W. Optimal scheduling of a taxi fleet with mixed electric and gasoline vehicles to service advance reservations. Transportation Research Part C: Emerging Technologies. 2018;**93**: 479-500

[74] He Y, Chowdhury M, Ma Y, Pisu P. Merging mobility and energy vision with hybrid electric vehicles and vehicle infrastructure integration. Energy Policy. 2012;**41**:599-609

[75] Mansour S, Harrabi I, Maier M, Joós G. Co-simulation study of performance trade-offs between centralised, distributed, and hybrid adaptive PEV charging algorithms. Computer Networks. 2015;**93**(Part 1): 153-165

[76] Hiermann G, Puchinger J, Ropke S, Hartl RF. The electric fleet size and mix vehicle routing problem with time windows and recharging stations. European Journal of Operational Research. 2016;**252**(3): 995-1018

[77] Naina PM, Rajamani H, Swarup KS. Modeling and simulation of virtual power plant in energy management system applications. In: 2017 7th International Conference on Power Systems (ICPS). 2017. pp. 392-397

[78] Park J, Chen Z, Kiliaris L, Kuang ML, Masrur MA, Phillips AM, et al. Intelligent vehicle power control based on machine learning of optimal control parameters and prediction of road type and traffic congestion. IEEE Transactions on Vehicular Technology. 2009;**58**(9):4741-4756

[79] Boriboonsomsin K, Barth MJ, Zhu W, Vu A. Eco-routing navigation system based on multisource historical and real-time traffic information. IEEE Transactions on Intelligent Transportation Systems. 2012;**13**(4): 1694-1704

[80] Bingham C, Walsh C, Carroll S. Impact of driving characteristics on electric vehicle energy consumption and range. IET Intelligent Transport Systems. 2012;**6**(1):29-35

[81] Lee J, Kang M-J, Park G-L. Battery consumption modeling for electric vehicles based on artificial neural networks. In: Lecture Notes in Computer Science. 2014. pp. 733-742

[82] Lee D-T, Shiah S-J, Lee C-M, Wang Y-C. State-of-charge estimation for electric scooters by using learning mechanisms. IEEE Transactions on Vehicular Technology. 2007;**56**(2): 544-556

[83] Regulation No 101 of the Economic Commission for Europe of the United Nations (UN/ECE). OJ L 158; 2007. pp. 34-105

[84] Clerc M, Kennedy J. The particle swarm-explosion, stability, and convergence in a multidimensional complex space. IEEE Transactions on Evolutionary Computation. 2002;**6**(1): 58-73

[85] Daramy-Williams E, Anable J, Grant-Muller S. A systematic review of the evidence on plug-in electric vehicle user experience. Transportation Research Part D: Transport and Environment. 2019;**71**:22-36

[86] Globisch J, Dütschke E, Schleich J. Acceptance of electric passenger cars in commercial fleets. Transportation Research Part A: Policy and Practice. 2018;**116**:122-129

[87] Masuch N, Keiser J, Lützenberger M, Albayrak S. Wind power-aware vehicle-to-grid algorithms for sustainable EV energy management systems. In: 2012 IEEE International on Electric Vehicle Conference (IEVC). 2012. pp. 1-7

[88] Valentine K, Temple W, Thomas RJ, Zhang KM. Relationship between wind power, electric vehicles and charger infrastructure in a two-settlement energy market. International Journal of Electrical Power & Energy Systems. 2016;**82**:225-232

[89] Strielkina A, Uzun D, Kharchenko V. Modelling of healthcare IoT using the queueing theory. In: 2017 9th IEEE International Conference on Intelligent Data Acquisition and Advanced Computing Systems: Technology and Applications (IDAACS). 2017. pp. 849-852

[90] Lakshmi C, Appa Iyer S. Application of queueing theory in health care: A literature review. Operations Research for Health Care. 2013;**2**(1): 25-39

[91] De Cauwer C, Van Mierlo J, Coosemans T. Energy consumption prediction for electric vehicles based on real-world data. Energies. 2015;**8**(8): 8573-8593

[92] Shankar R, Marco J. Method for estimating the energy consumption of electric vehicles and plug-in hybrid electric vehicles under real-world driving conditions. IET Intelligent Transport Systems. 2013;7(1):138-150

[93] Chan CC, Bouscayrol A, Chen K. Electric, hybrid, and fuel-cell vehicles: Architectures and modeling. IEEE Transactions on Vehicular Technology. 2010;**59**(2):589-598

[94] Holland JH. Adaptation in Natural and Artificial Systems: An Introductory Analysis with Applications to Biology, Control and Artificial Intelligence. Cambridge, MA, USA: MIT Press; 1992

[95] Goldberg DE. Genetic Algorithms in Search, Optimization and Machine Learning. 1st ed. Boston, MA, USA: Addison-Wesley Longman Publishing Co., Inc.; 1989